# Preface · 序

　　《悦食》是一本关于美食的书吗？可能吧，在一个"民以食为天"的国度里，美食是我们生活中最基本同时又是最重要的享乐之一。然而，随着时光变迁，在如今的消费时代中，食客们早已从单纯的"吃什么"上升至"哪里吃""如何吃"。因此这些承载美食本身的美食场所被食客和商家给予越来越多地关注和越来越多地期待。而书中收集的三十位设计师的四十二个案例，又为这些关注和期待提供了丰富的参照和无限的可能性。因此，《悦食》是一本关于为美食提供氛围空间的书。

　　此外，对于不同身份、不同群体、不同职业的人的理解中，"悦食"又会有着多样的诠释。

　　"悦食"对于食家而言，是一种进餐心情、一种享乐过程，是食家在独具匠心的设计与珍馐美味间游弋的味蕾记忆，是他们在设计师潜心营造的精美氛围中小酌一杯或大快朵颐时的享乐片段。在我们阅读作品"麻辣诱惑""清汤餐厅""金轩中餐厅"时，就能体会出这种食家心境。

　　"悦食"对于室内设计师而言，是一种创意方向、一种体验传达。设计餐厅与烹饪美食一样，要动用触觉、味觉、嗅觉、听觉、视觉，是设计师们通过对空间、色调、灯光、材质的全新调配，来营造这些配合美食的场所。而畅游在"大董金宝汇店""'外婆家'系列""所好轩"时，都能感受到设计师大厨级别的精湛厨艺。同时，我们也该适时地反思我们的设计，有无走进"过度设计"和"自娱设计"的误区。因为设计如美食，创新是最大的动力而自然才是最高的境界。

　　"悦食"对于餐厅业主而言，是一种最佳营运状态，是一种最直观的服务效果。在准确的市场调查和客层定位的前提下，提供与客户对接的文化氛围和用餐空间、细腻体贴的服务和独特优质的出品是一家餐厅被食客接纳认可的重要元素。正如行间老话，一厅、二堂、三厨房，而放在第一位的"厅"正是所讲的"用餐环境"的重要性。在作品"四季·恋""金屋国菜""SCENA意大利餐厅"的解读中，能让我们领悟到环境在引导消费、促进服务和配合出品中起到奇妙的作用。

　　所以《悦食》是一本关于感官体验、创意流程和经营模式的书。

　　另外我私下以为，《悦食》还是一本精妙的食谱，一本视觉盛宴的菜牌。食材丰富、口味多变、菜品丰盛。"悦食"还是阅读这本书的心态，以开放、包容和快乐的心情去解读这些美食空间就如享受满桌的美味佳肴。

　　好吧，祝您用餐愉快！

陈彬

2011年2月

悦食

*Delightful Restaurant*

*p6*

1. Bellagio餐厅
**B**ellagio Restaurant

*p14*

2. 沸腾鱼乡
**B**oiling Fish Village

*p20*

3. 箸香五角场店
**C**hopsticks Restaurant

*p28*

4. 大董金宝汇店
**D**aDong Roast Duck

*p40*

5. 多佐日式和风精致料理
**D**UO ZUO Japanese
Cuisine Restaurant

*p82*

11. 外婆家运动会
**G**randma's Kitchen

*p88*

12. 外婆家西溪店
**G**randma's Kitchen

*p94*

13. 外婆家湖滨店
**G**randma's Kitchen

*p98*

14. 外婆家万象城店
**G**randma's Kitchen

*p104*

15. 和府酒店中餐厅
**H**efu Hotel Restaurant

*p144*

21. 一口猪千禧大饭店
**M**R PIGGER
Restaurant

*p152*

22. 慕美餐坊
**M**u Mei Restaurant

*p158*

23. 新东方火锅店
**N**ew Oriental
Restaurant

*p162*

24. 新镇江酒家
**N**ew Zhen Jiang
Restaurant

*p168*

25. 兰亭别院
**O**rchid-Pavilion Court

*p206*

31. 上厢房餐厅
**S**hangxiangfang
Restaurant

*p212*

32. 麻辣诱惑梅龙镇店
**S**picy Temptation

*p220*

33. 麻辣诱惑淮海中路店
**S**picy Temptation

*p234*

34. 苏园酒店
**S**u Park Restaurant

*p240*

35. 牛公馆
**S**uper Noodle

*p280*

41. 烹大师烧肉达人
**Y**akiniku Master

*p286*

42. 一尊皇牛
**Y**izun Huangniu
Restaurant

*p292*

43. 中森名菜海岸城店
**Z**hong Sen Famous Food

# Contents · 目录

p46
p50
6. 所好轩
Favorite Pavilion

7. 金轩中餐厅
FLAIR Restaurant

8. 上岛西餐厅
Flowing Space

p58
p64
9. 北京丰沃德
FORWARD

p72
10. 四季·恋餐厅
Four Seasons Restaurant

p110
16. 金屋国菜
Jin Wu China Food

17. 荷畔餐厅
LAKEVIEW Restaurant

p118
p124
18. 刘家香餐厅
Liujiaxiang Restaurant

19. 厨房制造
Made In Kitchen

p130
p136
20. 厨房制造II
Made In Kitchen II

p174
p180
26. 完美生活
Perfect Life Restaurant

27. 清汤餐厅
Qingtang Restaurant

28. SCENA意大利餐厅
SCENA Restaurant

p186
p192
29. 膳福源
Shanfuyuan Restaurant

p200
30. 上海荣府
Shang Hai Rong Fu

p246
36. 唐会
Tang Hui Restaurant

37. 红馆餐厅
The Crimson Restaurant

p252
p258
38. 富田菊日本皇尚料理
Tomidagiku Restaurant

39. 南城记川菜馆
Traditional Sichuan Restaurant

p264
p270
40. 许仙楼
Xu Xian Restaurant

**吴伟宏：**

宏盟东方室内设计机构创意总监；

意大利米兰理工大学室内设计管理硕士；

IAI亚太建筑师与室内设计师联盟资深会员；

中国建筑装饰协会室内设计分会会员；

中国建筑装饰协会室内建筑师；

福建省建筑装饰行业 优秀青年设计师。

**近期主要荣膺：**

2010年，亚太设计双年大奖赛提名；亚太设计双年大奖赛最佳酒吧空间设计大奖提名。

2009年，荣获"大天杯"福建省室内与环境设计大奖赛公建类二等奖；荣获亚太室内设计大奖赛入围奖；亚太设计"中国风"大奖赛优秀奖。

2008年，荣获"星辉杯"福建省室内与环境设计大奖赛公建类二等奖。

2007年，荣获"辉煌杯"福建省室内与环境设计大奖赛公建类一等奖。

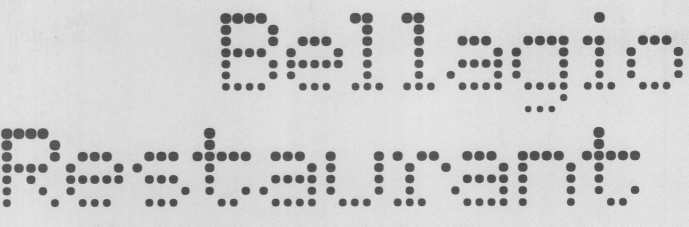

# Bellagio餐厅

项目面积：**400m²**
项目地点：福建厦门湖滨北路阳明楼1楼
设计公司：厦门宏盟东方室内设计机构
主案设计：吴伟宏
摄　　影：申强
客户名称：鹿港小镇集团

本案大型浅浮雕"波浪墙"的背景非常引人注目，它以一种安静的方式表现出漂移和游动，奇特舞台灯光让这整个空间充满梦幻的感觉，T台服务通道，就像一个走秀的舞台，不经意间体现设计师不拘一格、独具匠心的设计理念。

This huge bas-relief attractive by its 'Wave Wall' background. It shows the drifting and swimming in a quiet way. The fancy arena lighting makes the space full of dreamy feeling. T shape service channel which is just like a arena, shows the unlimited, spectacular designing idea unconsciously.

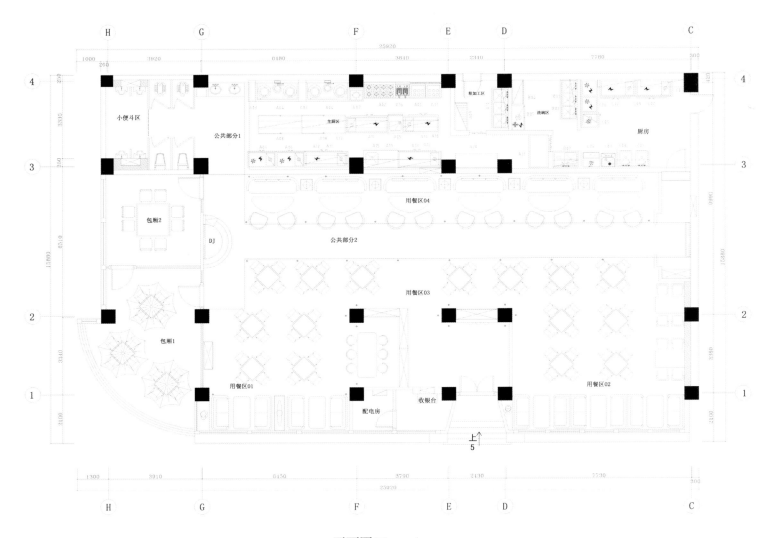

平面图/Floor plan

设计注重为使用者提供非正式交流空间，便利的休息设施充分体现了"以人为本"的理念。

The designer provide a informal communication space for users. The convenient leisure equipment shows the designing idea of 'People-Oriented'.

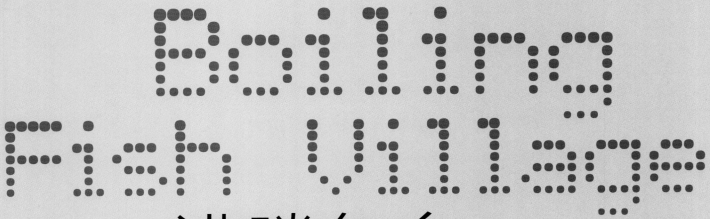

# Boiling Fish Village

# 沸腾鱼乡

## 北京世贸天阶店

项目地点：北京世贸天街北楼4层

概念设计：余青山

主持设计：徐宗

执行设计：韩宇

世贸天阶位于CBD商圈，沸腾鱼乡天阶店定位于广大追求品质的时尚人群。空间环境设计理念是在感受着时尚的同时，亦感受着时尚的精神，传承与进取。此空间场景整体统一为纯灰色基调，单纯而优雅，而ARDECO的装修风格又轩畅淋漓的讲述着细节与品质。

The Place in the CBD district, township days, boiling fish store located in order to pursue the quality of the majority of the fashion crowd. Space environment design experience is in fashion, they are also feeling the spirit of fashion, tradition and progress.The space scenes are pure uniform gray tone of the whole, simple and elegant, and ARDECO Cheong Hin decoration style and dripping with details about the quality.

处处有景观的开敞式空间设计，又将山、水、云、建筑这一概念化表达的自然环境意象贯穿全场，并从室内放大到了室外，消融了内外的心理界限，而当几件古老的中国艺术品置于其中，讲述一个古老的传说时，顿时打破了空间的"宁静"，整个空间瞬间激荡着鱼跃龙门这个古老传说所散发出的力量。中央鱼龙变化雕塑一冲云霄的鱼跃之势，呼之欲出亦风起云扬，无时无刻不带动着全场的沸腾之感。此时，已无需再分古今，整个空间讲述着传承与进取。纵观大环境整个空间的收与放、动与静、高与低、藏与露，所有的激撞与对比都轻盈的消于一缕清风之中，给人以心理及感观刺激的难忘瞬间。

Everywhere the landscape design of open-type space, in turn mountains, water, clouds, construction of the conceptual image of the expression of the natural environment through the audience, and from indoor to outdoor amplification, melting the inside and outside the mental boundaries, and when a few old placed in one of Chinese art, in an ancient legend about when, suddenly breaking space "quiet" moment of the entire space Yuelong Men stirring the ancient legend of the fish exudes strength. At this point, the entire space about the heritage and spirit. Throughout the entire space of the closed environment and release, dynamic and static, high and low, hidden and exposed, all the shock hit the light and contrast are among the consumer in the breeze, gives a psychological and memorable sensory stimuli moment.

在品赏着精致菜肴的同时，亦享受着优雅，感知着内涵，听这个空间所讲的故事，心理隐隐的暗合着这种精神，这就是对时尚的感悟，亦是沸腾鱼乡这个品牌一直追求并想给大家创造的全新就餐理念。

When tasting the exquisite dishes, we also enjoy an elegant, perceiving meaning, listening to stories told in this space, psychological faint coincide with this spirit. This is the perception of fashion has also been pursuing a restaurant and want to We created a new dining concept.

**马莹麟：**

室内设计师，毕业于上海纺织大学。

2001年成为昊元设计（国家乙级设计院）室内设计部的设计主管，2003年和伙伴共同创办林世装饰设计咨询有限公司，并担任设计总监。

## 近期主要荣膺：

2004-2005年度上海十大最具影响力室内设计师（林世）；

2004年度长三角最受欢迎室内设计师（中国房产协会）；

2005年度中国十佳住宅类室内设计师（中国建筑装饰协会）；

2007年度金外滩最佳色彩运用提名（美国《室内设计》中文版，中国建筑学会室内设计分会）；

2008年度金外滩最佳色彩运用提名（美国《室内设计》中文版，中国建筑学会室内设计分会）；

2009年，中国风 IAI2009亚太室内设计精英邀请赛新锐设计师。

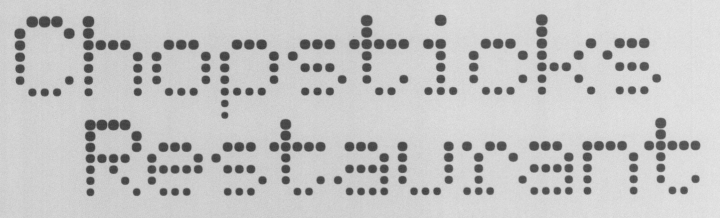

# 箸香五角场店

项目面积：*1100m²*

项目地址：上海市五角场

主案设计：马莹麟

设计公司：林世装饰设计咨询有限公司

本案的门头位于通道的顶端，所谓门厅其实是一条相对宽敞的走道。右边是水景，正对大门为迎宾台，左边可进入就餐区。在整个平面格局的中间醒目处，划出了较大一块比例作为吧台操作区。

整个前厅包括杯具、容器的清洗，果盘、冷饮、热茶、果汁的制作以及收银。吧台内还设有中岛，背景是到顶的半圆造型红酒柜（通过顶面镜子的折射变成了正圆），感觉就是一个大型的开敞式西式厨房。这样气氛即轻松又带有西方文化的感觉，很受顾客的喜爱。大型吧台其实亦把总平分成了两大部分：即左边是4-6人的长方桌（靠近景观窗），适合三五好友、情侣及小家庭；右边是8-10人的圆桌，可容纳100人左右的小型宴会区。整间餐厅有两间半包，6间VIP包间。

Head in the door of the case the top of the channel, the so-called hall is actually a relatively spacious walkways. The right is the water feature, is on the door for the reception table, on the left to enter the dining area. In the middle of the plane striking at the pattern, draw a large proportion of the operating area as a bar.
The entire lobby, including cups, containers, cleaning, fruit, cold drinks, hot tea, fruit juice production and the cash register. Background of the bar reach the peak of the semicircle shape of red wine cooler, feeling that a large Western-style open-style kitchen. This atmosphere is relaxed and with a sense of Western culture, popular customer favorite. Large-scale counter to divide the total into two parts: the left is one of the long table 4-6 for couples and small families use; the right is the 8-10 person round table, which can accommodate about 100 small banquet area. The whole restaurant has two semi-package, 6 VIP rooms.

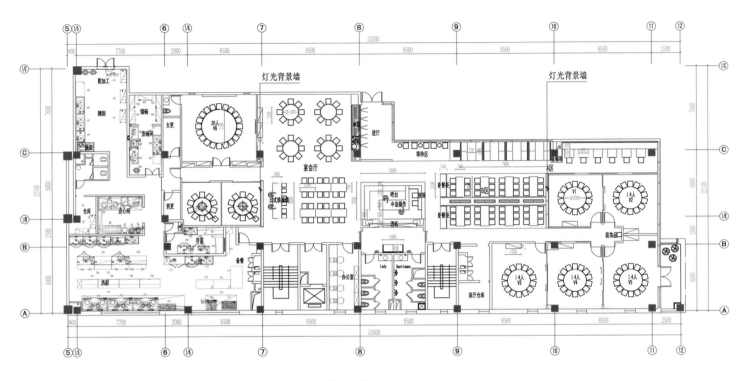

一楼平面图/Floor plan

"箸"的意思是筷子，带有很浓的中国文化。我们秉承一贯的设计思想：就是坚持中国文化演变至今的我们所理解的形态；不拘泥于材料、元素、甚至色彩，而在于其精神于意境。东方人的思想一直提倡平衡，人与人的平衡、人与自然的平衡，也可以将此理解为环保概念的雏形。

"Chopsticks" means chopsticks, with a strong sense of Chinese culture. We uphold the design idea: Stick to the shape of Chinese culture has evolved; not rigidly adhere to materials, elements, and even the color, but in their spiritual mood. Asian balance has been promoting the idea, people's balance, the balance between man and nature can also be understood as the prototype of the concept of environmental protection.

如今受国际大趋势的影响，我们会自然的考虑环保的因素，整间餐厅的材料以水泥乳胶漆为主，局部橡木饰面和少量的不锈钢镜面。其中水泥的颜色是一种天然的纯粹的灰，它所体现的质感有其独特的韵味。染黑的橡木保留原有的纹理，增添少许的厚重。不锈钢镜面的反射增加空间的进深。

Today, influenced by the international trend, we will naturally take into account environmental factors. Restaurant latex cement-based materials, local use of oak veneer and a small amount of stainless steel mirror. The color of cement which is purely a natural gray, reflects the unique charm. Black oak not only retain the original texture, but also add a little heavy. Stainless steel mirror reflections to increase the space into the deep.

顶面是局部吊顶，大面积的裸露设计。优点不用多说了，缺点是处理不好会让人觉得脏乱，白色更为明显。所以在走线、穿管、空调风口走向上要特别的小心，做到横平竖直。完成后效果就会有时尚感，白色亦会很好的起到灯光漫反射的作用，从而大大减少射灯的数量。当然还有其它辅助光源，我们多用LED灯管。

Top surface is a local ceiling, and large areas of bare design. Needless to say, the advantages and disadvantages are not handled properly will make people feel dirty, white will be more apparent. Therefore, alignment, wear control, air conditioning vents to be particularly careful on, so smooth vertical and horizontal. Complete results will have fashion sense, a good white will play the role of diffuse light, thus reducing the number of spotlights. Of course there are other secondary sources, we have multi-purpose LED lamp.

中式元素体现在格局、隔断、竹子图案的运用（灯罩）的软装饰品。设计师比较偏好汉代的风格，所以色系是黑、白、灰为主，一直贯彻至家居饰品。唯有灯具是暖色偏黄的像烛光的色泽。

但贯穿始终的还是现代与时尚，所有的元素经过拆解提炼重新排布。

Chinese elements reflected in the pattern, cut off, the utilization of bamboo pattern. Designers prefer the style of the Han Dynasty, so black, white, gray-based color, has been implementing to home accessories. Only lamps are warm yellowish color as the candle.
But still runs through the modern and fashionable, all the elements extracted from the new arrangement after dismantling.

# DaDong Roast Duck

# 大董金宝汇店

项目地址：北京市东城区金宝街金宝汇5层

概念设计：余青山

主持设计：徐宗

执行设计：张怀臣、郭科

主要材料：乳胶漆、中国黑烧毛石材、黑不锈钢镂
空雕刻、竹子、亚克力造型、山水玻璃

摄影：蒋晓维

中国水墨画是写意的，传神的，气韵生动之中是心灵的情态自由。笔与墨合，情与景合，情景交融。中国水墨画是诗性的，"诗有别趣，非关乎理"。一首词，几句诗都以投影的形式，贯穿整个空间。画中有诗，诗中有画，暗合中国古代文人之哲学。墨色美在单纯之中蕴涵了万物的光彩。大董空间犹如水墨写意之作，虽逸笔草草，却往往有笔外之笔、墨外之墨、意外之意，看上去漫不经心，但却耐人寻味。

Chinese ink painting is impressionistic, and spirit. Pen and ink together, Feelings and together, scenes. Chinese ink painting is a poetic, "not interested in poetry, non-related reasons." One word, several lines express all the projection in the form of, throughout the space. Painting in poetry, poetry paintings, ancient Chinese scholars coincide philosophy. Ink into the United States implies the simple glory of all things. Board space is like a large ink freehand to make, though Yat pen hastily, but often have a pen outside the pen, the ink outside the ink, meaning accident and looked casual, but intriguing.

浮云般的写意花朵，在大厅区上任意绽放。
Cloud-like impressionistic flowers, blooming in the lobby area on the arbitrary.

包间运用取情取景的建构方式，墙面被栏杆、移门所替代。可开可合，或倚栏凝想。
Rooms used to take the situation into the construction of framing, the walls are bars, sliding door replaced. Can open, you can also shut down.

大董空间是水墨意境与文人情感的载体，心境可以在拙然的古琴声中得以平静，情怀可以在平淡自然的空间中得以袒露。水墨肇自然之性，知黑守白。在最单纯的黑白两色空间里，彩蝶翩翩，当你开启房门的那一刻，她们簇拥着你，迷恋着你，带着那金色的光辉，似墨笔不经意间沾染了一缕霞光，如雨后初晴，又或者是暮夏夕阳。

Artistic and literary space is emotional ink carrier, in the humble state of mind can be calm in the natural sound of the piano, simple and natural feelings can be bare space.In the most simple black and white space, butterflies fly, when you open the door the moment, they surround you, crush on you, with a brilliant golden, pen inadvertently contaminated with a ray of rays. As the sunny day after rain, or a summer sunset.

竹，虚心文雅，挺拔秀丽。鹤，超凡脱俗，高贵优雅。两者都与中国文人有着不解之缘，作为贯穿空间的符号，我们更将它们赋予全新的表达形式，或以光影投于墙面之上，如明月般浮于水景之中；或以浓墨粗笔的剪影形式闪现于竹林之中。灵秀慧石般的洗手盆，涓涓的水声，勾画出清幽宜人，潇洒淡泊的男卫生间。

桃花三两枝，曼妙的心境，女卫生间那悠然的古典情韵，婀娜的向我们缓缓走来，温润的白瓷桃花，委婉的散发着幽香，而思绪却早已徜徉在花飞花谢的幻想之中。

Chinese scholars like bamboo and cranes, which are space symbols. We will give them a new form of expression, or to cast shadows on the wall above the moon as if floating on the water features being; or thick pen thick and black silhouette of bamboo being in the form of flash. Stone wash basin, trickling sound of water, described a pleasant, handsome male toilet.

Beautiful peach blossoms, graceful mood, female bathroom that classic Sentiment leisurely, slowly coming to us, moist white porcelain, the distribution of fragrance, while the mind is already wandering in the flowers fade fantasies.

**王俊钦：**

　　睿智匯设计总经理兼总设计师。中国建筑学会室内设计分会，会员；中国照明学会，高级会员。出生于中国台湾，毕业于中国工商管理学院建筑系(台湾)，在娱乐空间、商业空间、餐饮空间、会所空间、办公医疗、豪宅别墅设计持有专长20多年来他因地制宜，"多佐餐厅"，"麦乐迪KTV系列"，"普罗旺斯豪宅"，为他赢得了2010中国室内设计师年度封面人物大奖；2010年金堂奖年度娱乐空间设计十佳，金堂奖年度别墅设计十佳和金盘奖年度最佳样板房，金堂奖年度餐饮优秀作品，奠定了他在中国设计界的地位，被誉为"理性和感性双全的设计师"。

**近期主要荣膺：**

　　2009年，中国室内设计师年度提名封面人物。

　　2009-2010年度，中国室内设计百强人物；中国国际设计艺术博览会商业空间类一等奖；西顿照明杯年度最具人气案例入围奖。

　　2009年度，中国最具价值的室内设计企业；中国创意产业高成长100强。

　　2010年，中国室内设计师年度封面人物。

　　2010德国ZUMTOBEL奥德堡照明室内照明设计银奖；

　　2010金外滩，"最佳餐厨空间优胜奖"；

　　2010金堂奖，年度十佳娱乐空间设计作品；

　　2010金堂奖，年度十佳别墅设计作品；

　　2010金堂奖，年度餐饮优秀作品；

　　2010金盘奖，年度最佳样板房。

# DUO ZUO Japanese Cuisine
# 多佐日式和风精致料理

项目面积：*1300m²*

项目地点：北京市双井桥

设计公司：睿智匯设计公司

设计：王俊钦 彭晴

主要材料：黑色镀锌铁板激光雕刻喷涂、茶色玻璃、金属帘、茶镜、橡木饰面板

北京多佐日式精致料理餐厅坐落于北京CBD黄金商业圈，占地面积1300平方米。这里的设计充分把日本料理文化精髓用形式感，材料美，色彩新表达出来，通过运用"意境"古典美学的独特范畴，深深的植根于日本的文艺与中国哲学土壤中。讲求的不仅在于"形"，更在于"形外之象"，即审美主体的"言有尽而意无穷"。日本的设计更多以日本传统的空灵虚无思想为根底，带有日本自古以来的清愁冷艳的色调，追求其中浮现的优美和冷艳的感情世界。

Restaurant is located in Beijing CBD business circle of gold, covers an area of 1,300 square meters. This design fully to express the essence of Japanese culture, through the use of "mood" of the unique areas of classical aesthetics, deeply rooted in Japanese literature and Chinese philosophy soil. Not only is the emphasis on "form", but also because the "spirit", that is the aesthetic subject "mood." Japanese design more traditional thinking of the Japanese origin, with the colors of Japan since ancient times, the pursuit of one of the feelings of the world.

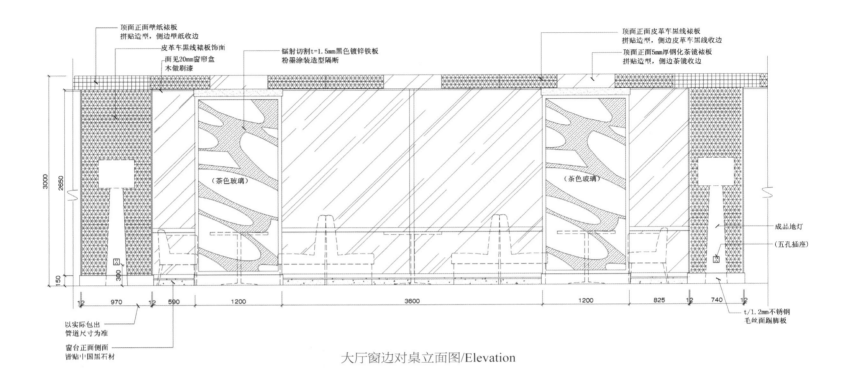

顶面正面壁纸裱板
拼贴造型，侧边壁纸收边
皮革车黑线裱板饰面
面见20mm窗帘盒
木做刷漆

镭射切割t=1.5mm黑色镀锌铁板
粉墨涂装造型隔断

顶面正面皮革车黑线裱板
拼贴造型，侧边皮革车黑线收边
顶面正面5mm厚钢化茶镜裱板
拼贴造型，侧边茶镜收边

（茶色玻璃）

（茶色玻璃）

成品地灯

（五孔插座）

以实际包出
管道尺寸为准

窗台正面侧面
皆贴中国黑石材

t/1.2mm不锈钢
毛丝面踢脚板

3000
2650
150
300

12  970  12  590  1200  3600  1200  825  12  740  12

大厅窗边对桌立面图/Elevation

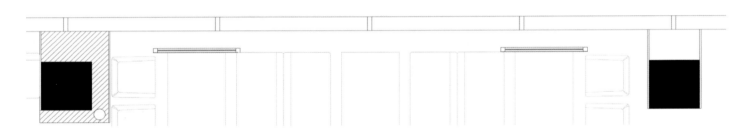

大厅窗边对桌平面图/Elevation

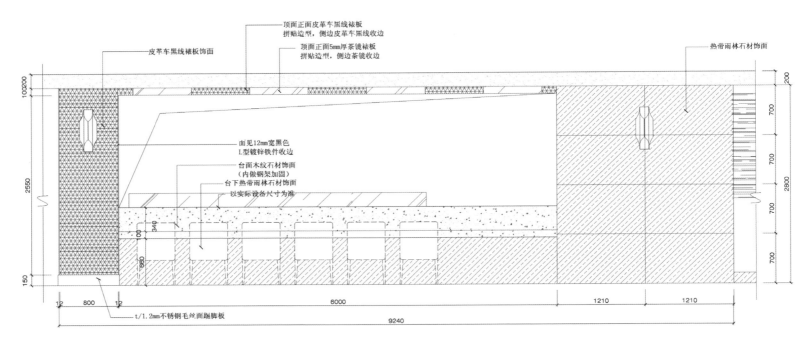

皮革车黑线裱板饰面

顶面正面皮革车黑线裱板
拼贴造型，侧边皮革车黑线收边

顶面正面5mm厚茶镜裱板
拼贴造型，侧边茶镜收边

热带雨林石材饰面

面见12mm宽黑色
L型镀锌铁件收边

台面木纹石材饰面
（内做钢架加固）
台下带热带雨林石材饰面
以实际设备尺寸为准

t/1.2mm不锈钢毛丝面踢脚板

刺身吧立面图/Elevation

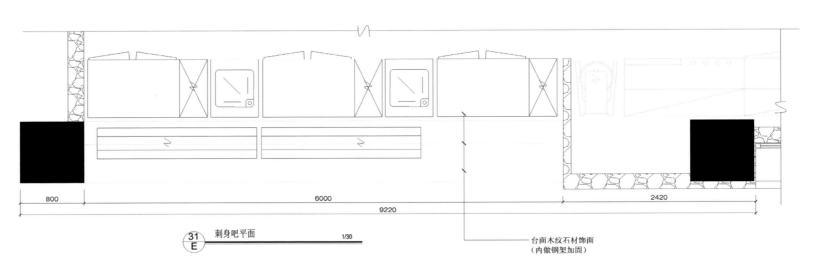

③①
Ⓔ   刺身吧平面          1/30

台面木纹石材饰面
（内做钢架加固）

刺身吧平面图/Elevation

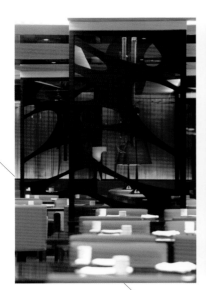

由于受到不同地域环境的影响，采取了因地制宜的设计概念，整体运用了暖色调，给人一种冬季中温暖般感觉，避免传达一种冷清的意境。由此，决定了有精挑细选的食材、华丽的摆盘，在整体装饰上也力求区别于传统的日式料理风格。

As a result of the environmental impact of the different regions, to take the design concept of local conditions, the overall use of warm, giving a warm feeling. As a result, determines a carefully selected ingredients, gorgeous tableware, decoration and strives in the whole different from the traditional style of Japanese cuisine.

陈彬：

武汉理工大学艺术设计学院副教授

中国建筑学会室内设计分会（CIID）会员

中国建筑装饰协会会员及设计委员会委员

国际室内建筑师/设计师联盟（IFI）专业会员

亚太建筑师与室内设计师联盟（IAI）理事

大木（湖北）后象设计顾问机构 设计主持

## 近期主要荣膺：

作品入选ANDREW MARTIN 2010年度国际室内设计大奖；

2010年，亚太室内设计双年大奖赛-评审团特别大奖；

2010年度，广州国际设计周"金堂奖"年度十佳餐饮空间设计；

2009年度，APIDA第十七届亚太区室内设计大奖银奖；

2009年度，作品入选德国iF2009中国设计大奖；

2009年度，广州国际设计周"金羊奖"中国十大室内设计师；

2009年度，广州国际设计周"金堂奖"餐饮酒吧类中国十大设计师；

2009年度，Interior Design China "酒店餐厅类最佳设计"奖；

2009年度，金指环全球室内设计大赛优秀奖。

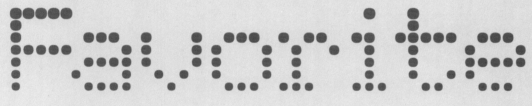

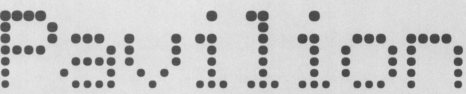

# 所好轩

项目面积：*518m²*

项目地点：武汉西北湖

完成设计时间：*2009年11月*

设计公司：大木设计中国(湖北)后象设计顾问机构

主持设计师：陈彬

参与设计师：李健、傅晟、周翔、杨慧

摄影：吴辉

《孟子告子》中曾说"食、色、性也",形象表达出中国古人对美食、美景的所好、所恋。曹雪芹为好友题写书斋篇,其中书"色宜少年,食宜饥腹",以中国文人特有的方式,体验出美景美食的境界,书斋名曰"所好轩"。

餐厅位于中国武汉,是一家大型餐饮店的店中店,设计师以古文人的"食色宣言"为灵感,用色彩和器物营造出一种"花鸟鱼虫"的中式休闲意境。

In *Book of Mencius*, Gaozi first pointed out that food and sex are the two basic necessities for mankind and longing for the two is the nature of mankind. This famous saying vividly illustrates that the Chinese ancient people had been pursuing favorite food and beautiful appearance. Cao Xueqin, a famous novel writer in Qing Dynasty, ever wrote in an article for the study room of his good friend that paying much attention to sex in youth is better than in any other life stage and food is more favorite and delicious especially when one is hungry, which manifests men of letters have been longing for delicious food and attractive beauty in their unique way. The study room was named Favorite Pavilion. Our restaurant is actually named after this study room.

It is located within a big chain restaurant in Wuhan, China. Deriving inspiration from the ancient people's *Declaration of Food*, our designers have successfully created a Chinese-style leisure atmosphere integrating both static and dynamic touches such as flowers, birds, fish and insects with the help of colors and objects.

平面图/Plan

**梁志天：**

香港十大顶尖设计师之一；

香港大学建筑学学士；

香港大学城市规划硕士。

1997年，他创立了梁志天建筑师有限公司及梁志天设计有限公司；

2000年，于上海设立梁志天设计咨询(上海)有限公司。

**近期主要荣膺：**

2006年，梁氏已六度获得素有室内设计奥斯卡之称的AndrewMartinInternationalAwards殷选为全球著名室内设计师之一；

2002年，香港薄扶林道宝翠园私人会所获得"2002亚太区室内设计大奖"私人会所类别组冠军，深圳观澜豪园获得住宅类别组优异奖；

2002年，香港山顶加列山道的赛诗阁、深圳观澜豪园等多个住宅设计项目获得"HKDA Design02 Show—Excellence Award"。

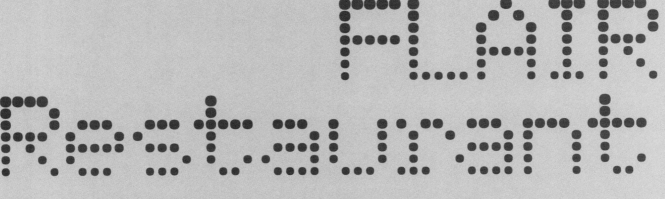

# 金轩中餐厅

项目地点：上海浦东丽思卡尔顿酒店的53和55层

设计师：梁志天

竣工时间：2010年6月

摄影：申强

金轩中餐厅是香港的著名设计师梁志天先生(Mr. Steve Leung)的得意之作，它位于酒店的53和55层，室内装潢和沪城美景相互辉映，趣意盎然。金轩绝对是城中社交名流、商务显达的不二之选。菜谱以地道粤菜为主，伴以各省精选名菜。而精妙的地方更在于其对茶道的讲究，这里根据客人点选的菜式，提供最好的茶叶搭配，并全以热壶侍奉。

This restaurant is the finest creations of Mr. Steve Leung,who is the famous designers from Hong Kong. It is located on 53 and 55 layers, upholstery and outdoor views embraced. Restaurant is definitely the city social and business premises of choice.The main menu to authentic Cantonese cuisine, accompanied by the provinces selected dishes. The place is more subtle in its stress on the tea ceremony, where guests select dishes according to provide them with the best tea.

金轩大厅可同时容纳52位宾客就餐。6个独立私密包间则可容纳一共64位宾客。2个豪华贵宾厅可供40位宾客宴客用膳。餐厅还特设中国茶廊，可供18位客人品茗。

Hall can accommodate 52 guests. Six independent private rooms can accommodate a total of 64 guests. Two luxurious VIP room for 40 guests banquet meal. The restaurant also has special gallery of Chinese tea, for 18 guests.

李川道：

东易国际（福州）创意中心 首席设计师；

室内建筑师；

全国优秀设计师；

中国建筑学会室内设计分会（CIID）会员；

国际室内建筑师/设计师联盟（IFI）会员。

# Flowing Space

# 上岛西餐厅

项目面积：780m²

项目地址：福建

竣工时间：2010年5月22日

主案设计：李川道

软装设计：陈立惠

摄影：申强

本案为上岛西餐厅，设计师以其商业定位为基点出发，结合自己的设计创意，勾画了一处时尚与浪漫并具的空间氛围。

This case of Shangdao Western-style Restaurant is a design of breathtaking atmosphere of fashion and romance, embodying designers' starting point of commercial positioning and combination of their vivid design ideas.

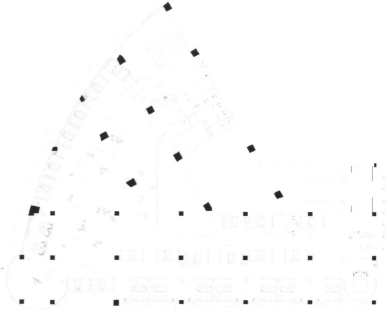

平面图/Floor plan

巧妙的光影处理是设计师在此空间里所要表达的首要途径。入门处，纯洁的白玫瑰木饰成为灯源的最佳佐料，镜面的设置使得灯光层次感更丰富，制造出浪漫的第一视觉感观。餐厅内，一个个夜光杯散射出绚烂的灯光，营造出焕彩优雅的氛围，整个空间如同沐浴在温暖的深海之中，浪漫的气息宛若鱼群在空间中肆意窜动。点状的灯光从天花板倾泻而下，吧台处晶莹的泡泡在闪烁，光影虚实间，美在流动。

Consummate skills of dealing with light and shadow are the desighers' principal way of expression. At the entrance,the pure decorative wood of white roses acts as the best ingredient of light source. The settings of mirror surface capture the first visual sensibility of romance that adds depths to lighting. Inside the restaurant, you will enjoy feasting your eyes on the dazzling lights, given out by luminous wine glasses ,which set up a blaze of colors that flame and flicker across a background of eleglance. The whole space is immersed in the sea of warmth, in which romance seems like fish swimming leisurely and lightheartedly. With the pouring down of punctate light, crystal clear foam makes its way across the bar counter ,leaving nothing but silent beauty in its wake.

此外，设计师对于空间的区域过渡划分也十分巧妙，采用了有趣的隔断手法，飘扬的曲线围合及环绕的夜光杯既打破空间秩序，又塑造了秩序，增添了空间的时尚质感，起起伏伏的节奏犹如悠扬的琴声四处涌动。同时，设计师也巧妙地将低碳概念融入空间创作之中，大部分灯光均为低压照明，既节省了商业成本，也是出于空间环境的深层思考。

Furthermore, by using ingenious designs of spatial area transition and employing artful skills of partition and curve enclosing, the designers achieve their goals to break the traditional spatial order and  to convey modern society peculiar rhythm which sounds like hauntingly beautiful music pouring out of piano. Meanwhile, designers masterly apply low-carbon conception to spatial creation and then tactfully bring low-voltage lighting into full play. In addition to saving commercial cost, low-voltage lighting also takes spatial environment into consideration.

**迫庆一郎：**

1994年，毕业于东京工业大学；

1996年，东京工业大学研究生毕业；

1996年-2004年，就职于山本理显设计工场；

2004年，成立SAKO建筑设计工社（中国北京）；

2004年，主持东方设计公社；

2004年-2005年，作为日本文化厅的外派艺术家赴哥伦比亚大学担任客座研究员。

## 近期主要荣膺：

2006年，商业地产奖（中国）：「天津万花景」；亚洲（公寓）金奖（中国）：「济南节奏」；日本商业空间协会赏（日本）审查者奖（桥本夕纪夫奖）「北京喜彪」（CIBOL）：银奖「北京蒲蒲兰绘本馆」，入选「北京芬理希梦2」。

2007年，日本商业空间协会赏（日本）-入选「杭州浪漫一身」；中国住户设计赏，创新十强楼盘-优秀赏（中国）：「北京冲击」；绿色／人文／科学地产，科技先锋-优秀赏（中国）：「北京冲击」；鹰牌陶瓷杯全国室内设计大赛商用空间类1等奖：「北京蒲蒲兰绘本馆」；JCD（商环境设计家协会）设计奖（日本）-银奖「杭州浪漫一身2」，入选「巴塞罗那IMAGINARIUM」。

2008年，第五届（2007）现代装修国际传媒奖（中国）：年度杰出设计师奖；鹰牌陶瓷杯全国室内设计大赛商用空间类 1等奖：「北京蒲蒲兰绘本馆」；日本商业空间协会赏(日本)，银奖：「杭州浪漫一身2」，入选：「巴塞罗那IMAGINARIUM」。

2008年度，日本GOOD DESIGN奖；中国第二届十佳配饰设计师奖

2009年，EuroShop RetailDesign Award 2009 .One of Three Best Stores Worldwide：「杭州浪漫一身2」．

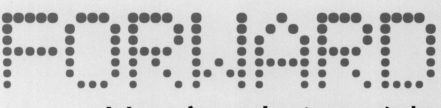

# 北京丰沃德

项目面积：*1050m²*

项目地点：中国北京市

设计：*SAKO*建筑设计工社

设计师：迫庆一郎

摄影：*Misae HIROMATSU*

蜿蜒的方格隔断隔出来的中餐厅——木板编织成方格图案，并按水平垂直方向将编织好的木板弯曲成波浪起伏的连漪的形状，从而形成了柔软和力量兼备的刚柔并济的隔断。并由此延展而成开放的窗边位置，中间区域以及里侧的封闭式包间群这三类空间。

Separated from the winding square off in the restaurant - wood woven grid pattern, horizontal and vertical direction according to weave a good wood bent into the shape of the ripple waves, creating a power of both soft and rigid and cut off the soft economy. And thus extend the window from opening position, the middle side of the closed areas and rooms in three categories of space groups.

餐厅入口，通过使用玻璃、水、镜面以及镜面不锈钢制作的水槽和展架，营造出招徕顾客的一个独特的入口空间。

The restaurant entrance, through the use of glass, water, mirror and stainless steel sink and mirror display rack, create a unique entrance space to attract customers.

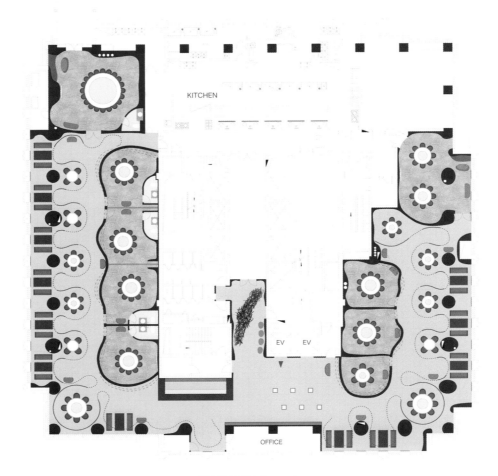

一楼平面图/Floor plan

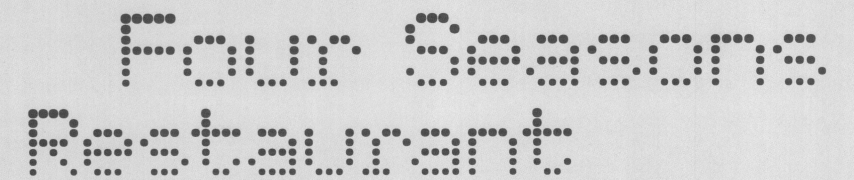

# Four Seasons Restaurant

# 四季·恋餐厅

项目面积：*3500 m²*
项目地址：湖北武汉
设计师：张猛
摄影：吴辉

本案的光谷店可说是四季恋的旗舰店，满足了一大批身在武昌的"四季恋迷"的需求。光谷店的规模最大，装修更为豪华——但绝不是拒人千里之外的那种豪华。

Optics Valley of the case can be said that Four Seasons store flagship store to meet a large number are in Wuchang's "Four Seasons fans"needs. Optics Valley of the largest shop, decorated --- but it is far more luxurious thousands of miles away to resist the kind of luxury.

餐厅几乎是纯白的风格，曲线与弧线交错的前卫设计。你走进来后的第一感觉就是"好像自己在拍电影"。要是坐在其中吃饭，更是一种享受。除了梦幻的环境、好吃的菜、高级的服务以外，甚至还有现场钢琴演奏。

The style of the restaurant is almost pure white, curves and arcs cross the avant-garde design. You come in the first feeling is "like myself in a movie. " If sitting in one meal, it is a pleasure. In addition to fantasy environment, delicious food, superior service outside, and it even has live piano performances.

**孙云:**

　　1992年，毕业于上海戏剧学院舞台空间设计专业。

　　杭州内建筑设计有限公司合伙人、设计总监；

　　高级室内建筑师。

**沈雷:**

　　1992年，毕业于中国美术学院环境艺术系；

　　2001年，英国爱丁堡艺术学院设计硕士毕业；

　　2002年-2004年，担任ID+C杂志执行主编。

　　杭州内建筑设计有限公司合伙人、设计总监；

　　高级室内建筑师，英国注册建筑师。

**姚路:**

　　1992年，毕业于浙江丝绸学院；

　　2000年，创办大金商业展示设计有限公司，从事商业终端展示设计工作。

　　杭州内建筑设计有限公司合伙人、设计总监；

　　高级室内建筑师，品牌策划师。

# 外婆家运动会

项目面积：*1000m$^2$*
项目地址：杭州古水街
设计单位：内建筑设计事务所
主要材料：木材、石料、钢
摄影：申强

因为曾经从事过体育运动的业主与体育界颇有深厚的渊源，开一家以运动为主题的餐馆也就应运而生。以运动为切入点的主题设计本身就具备了亲和的特质，充满了趣味与参与感，吸引着人们迅速地融入其中。

Because the owners had engaged in sports has its roots in the sports sector, restaurants to sports as the theme. The theme for the entry point to the design movement itself has the characteristics of pro-and full of fun, drawing people quickly into one.

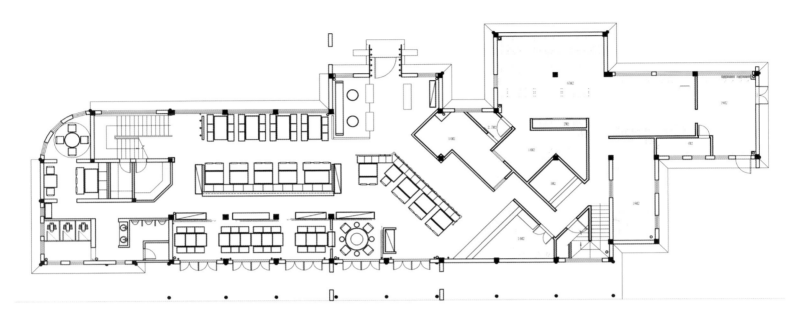

一楼平面图/Floor plan

餐厅分为上、下两层，就餐区域基本采用开放式格局以保持动线流畅，人流可以自然地进入各功能区。

Restaurant is divided into two layers, the basic use of an open dining area to maintain a steady pattern. Guests can access all functions of natural areas.

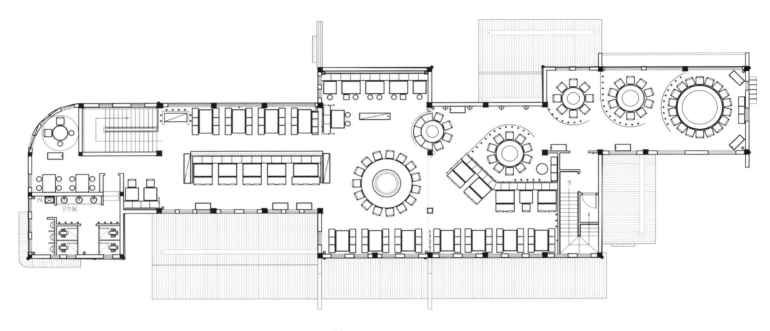

二楼平面图/2nd floor plan

为满足不同顾客对就餐私密性
的不同需求，下层就餐区以陈列
柜区隔出公共及半私密就餐区
域，上层则以帷幔圈围出一些
更具私密感的包间形式就餐区
域。而空间中大量的留白设计
则让"运动"这一主题得以充分
展现和延伸。

To meet different customer
privacy on the needs of
eating, the lower dining area
separated from the dining
area to showcase the upper
to the curtain surrounded
the dining area with a
sense of privacy. And a lot
of blank space in design
is to "exercise" the subject
can be fully displayed and
extension.

### 孙云：

1992年，毕业于上海戏剧学院舞台空间设计专业。

杭州内建筑设计有限公司合伙人、设计总监；

高级室内建筑师。

### 沈雷：

1992年，毕业于中国美术学院环境艺术系；

2001年，英国爱丁堡艺术学院设计硕士毕业；

2002年-2004年，担任ID+C杂志执行主编。

杭州内建筑设计有限公司合伙人、设计总监；

高级室内建筑师，英国注册建筑师。

### 姚路：

1992年，毕业于浙江丝绸学院；

2000年，创办大金商业展示设计有限公司，从事商业终端展示设计工作。

杭州内建筑设计有限公司合伙人、设计总监；

高级室内建筑师，品牌策划师。

# 外婆家西溪店

项目面积：*2000m²*
项目地址：杭州紫荆花路
设计单位：内建筑设计事务所
主要材料：木材、石料、玻璃、钢
摄影：申强

质感时尚是空间的主体语言，天花上的手写菜单和大幅面的张拉膜图象成为最鲜明的表达语素，附着于空间表皮之上，带来新鲜的视觉冲击。而旧物的点缀穿插则给空间带来了丰富的面相。

Space is the main language of fashion, handwritten menu on the ceiling and large format images of the most distinctive expression of language, attached to the space above the skin, bringing a fresh visual impact. Interspersed with the old objects are decorated to bring a wealth of face space.

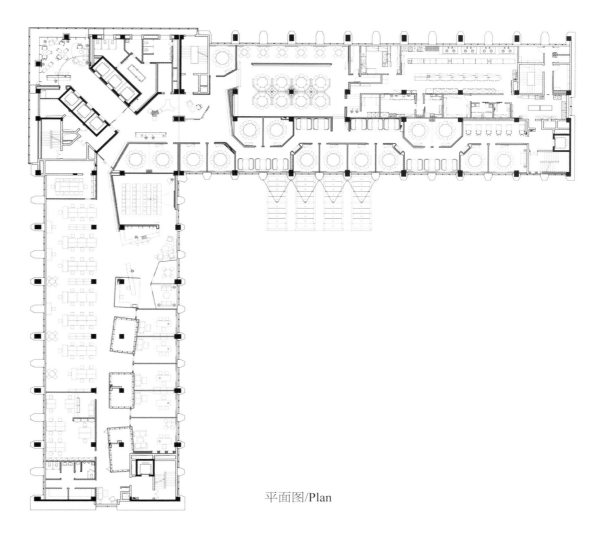

平面图/Plan

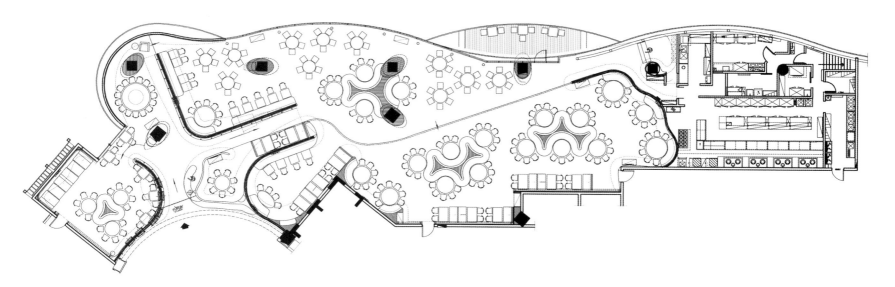

平面图/Plan

张驰：

　　中国建筑学会重庆室内分会副秘书长（中国建筑学会）；

　　中国装饰协会设计委员会委员（中国建筑装饰协会）；

　　中国百佳优秀室内建筑师（中国建筑学会）；

　　中国商业空间设计五十强设计师（中国装饰协会）；

　　中国杰出青年设计师（中国装饰协会）；

　　中国室内设计十大年度封面人物（提名）（中国建筑学会、美国《室内设计》中文版）；

　　中国工商联合会会员（中国工商联合会）；

　　西南精英室内设计师（中国建筑学会）；

　　四川美术学院室内设计专业特聘教师（四川美术学院）；

　　重庆装饰协会设计委委员（重庆装饰协会）；

　　MDC美盛酒店设计-张驰室内建筑师事务所负责人。

近期主要荣膺：

　　2007年，中国室内设计大奖赛及第二届世界室内建筑师联盟大奖赛佳作奖；

　　2009年，中国"金外滩"最佳照明设计奖；

　　2010年，中国"金外滩"最佳酒店设计优秀奖；

　　2010年，美国"酒店"杂志最佳酒店餐饮奖。

# Hefu Hotel Restaurant

# 和府酒店中餐厅

项目面积：*2500m²*
项目地址：重庆市江北区红锦大道2号
主案设计：张驰
主要材料：黑金花石材、墙布、墙纸、地毯
摄影：张驰、吴辉

从前，人们认为中式风格就是奢华。在世界文化中，中国的浪漫与富足在经由马可·波罗近乎崇拜的笔触描下：中国，是一个由宫殿和宝塔组成的国度,异国的风情和令人垂涎的财富；名贵的香料与华丽堂皇的宅院；精美的细木家俱；绚丽的丝绸和玲珑的瓷器。由此，浪漫和富足成为中式风格的不二标签。

我们常常会不由自主地被这种华丽所诱惑，以绚丽的色彩营造奢华的氛围，但同时我们也希望回归道家与儒家提倡的简朴生活原则。强调平衡、规律与协调性。指出简单与和谐对于生活的重要意义。在这种思潮下，简约的新中式成为这个酒店餐饮项目的注解。线条优雅简洁，结构精密合理，比例完美协调，功能齐备高效。它努力以东方哲学为内涵，试图将古老的中国艺术风格与现代极简主义结合。希借此融入全球室内设计的新潮中。

In the past, people think luxury is the Chinese style. Culture in the world, China's romance with the wealthy in the pen by Marco Polo: China is a country composed of palaces and pagodas, exotic customs and gave birth to the wealth of people down; rare spices and razzmatazz of the house; fine fine wood furniture; silk and porcelain. As a result, romantic and rich style of a Chinese label.
We often can not help being tempted by this gorgeous, with brilliant colors to create a luxurious atmosphere, but we also hope to promote the return of the Taoist and Confucian principles of simple life. Stress balance, and coordination of law. Pointed out that the simple and harmonious life for the significance. In this thought, the simplicity of the new Chinese restaurant projects a comment to this hotel. Elegant lines and simple, precise and reasonable structure, the proportion of perfect coordination, full-featured and efficient. It is hard to Eastern philosophy as the content, trying to ancient Chinese art style combined with modern minimalism. Hope to integrate into the global fashion in interior design.

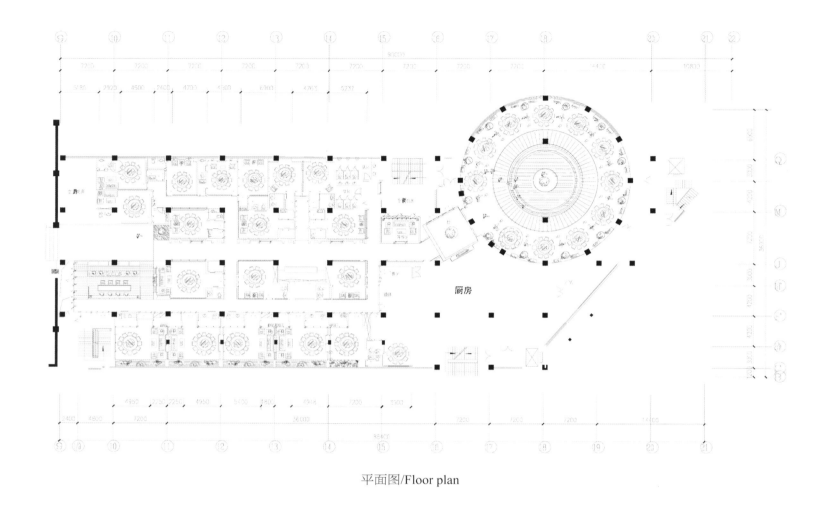

平面图/Floor plan

**林文格：**

L&A文格空间设计顾问公司创办人，创意总监；

高级室内建筑师；

全国百名优秀室内建筑师；

IFDA国际室内装饰设计协会理事；

ICAD国际A级职业景观设计师；

中国建筑学会室内设计分会第三专业委员会委员；

香港室内设计协会中国深圳代表处委员；

意大利米兰理工设计学院室内设计硕士。

**近期主要荣膺：**

2010年，英国Andrew Martin 设计大奖；

2009年，美国Hospitality Design Awards 酒店空间设计大赛最高荣誉winner大奖；

2009年，荣获中国建筑学会"中国室内设计二十年二十人"荣誉称号；

2009年，荣获中国建筑学会"中国杰出室内建筑师"荣誉称号；

2009世界酒店"五洲钻石奖"-"最佳设计师"；

APIDA第11届亚太区室内设计大赛餐馆酒吧类别冠军奖；

APIDA第十四届亚太区室内设计大赛酒店类别铜奖。

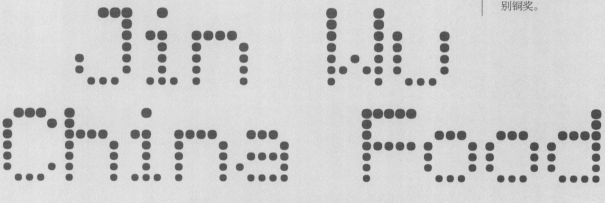

# 金屋国菜

项目地址：深圳福田区福中三路1006号诺德金融中心1楼

设计公司：文格空间-设计顾问（深圳）公司

设计师：林文格

就餐厅名字而言，颇有些让人浮想联翩的意境。金色，在传统风俗中是帝王之色，被视为超俗之色，是皇家专用色彩象征，有着任何色彩都无法代替的贵气与华丽。传承中国历代文化精髓，设计师的脑海中呈现出了一系列闪亮的元素：金算盘、金箔、紫铜镂刻图纹……于是设计构思也便随着想象的丰富而一点点的呈现出来。

Name for the restaurant, the mood quite some appeals to the imagination. Gold is king in the traditional customs of the color, is considered ultra-custom color, and is a symbol of the Royal color, with any color can not be replaced by extravagance and splendor. The essence of China's ancient cultural heritage, the designers of the mind showing a series of flash elements: gold abacus, gold, copper engraving pattern, ... ... so design idea will be with the imagination and a little bit of showing it.

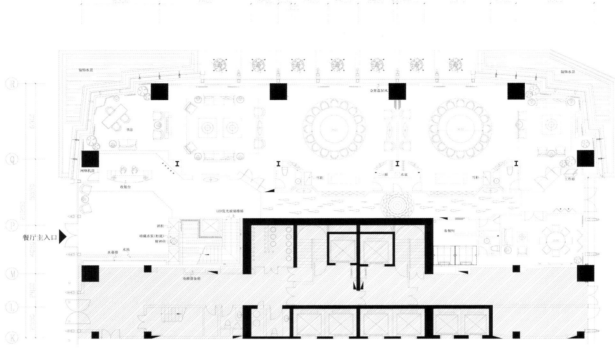

平面图/Plan

色彩上渲染奢华: 为更好的演绎餐厅所表达的设计理念，巧妙运用了金黄色特有的贵族气息，以不同的材料、造型及其它元素，展示在空间的每一个角落。黑色给人的感觉是高贵、神秘、庄重。设计师以黑色基底衬体，与金色的相容镶嵌营造出让人意想不到的空间效果。

Rendered in color, luxury: for better interpretation of the design concept expressed in the restaurant, clever use of the unique golden nobility to different materials, shapes and other elements, displayed in every corner of space. Black gives the feeling of noble, mysterious, solemn. Designer use black base underlay the body, and the compatibility of gold to create a mosaic effect of unexpected space.

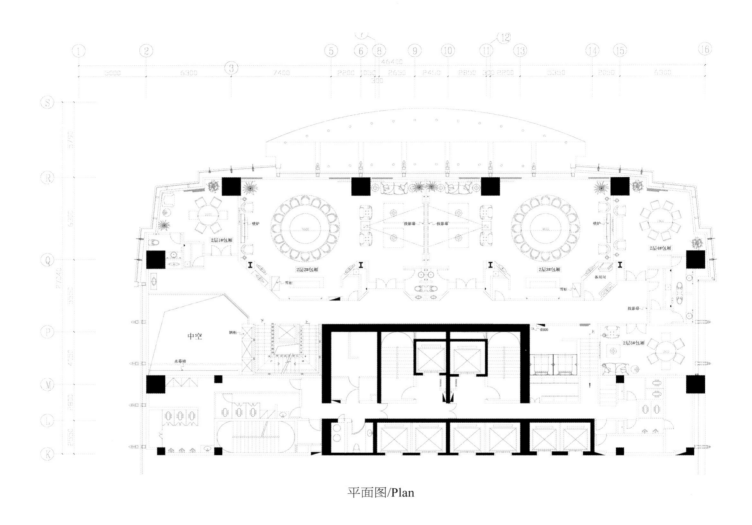

平面图/Plan

灯光上营造雅致：在就餐环境中，灯光是调节情绪、气氛的重要手段。在本案中，通过镂刻的紫铜吊顶，灯光光线被柔和的放射出来，光与影的巧妙结合，在整个天花上演绎着一曲曲律动的舞姿，尽显张力。

Lights on to create elegant: In the dining environment lighting is an important means to adjust the mood and the atmosphere. The present case, by engraving the copper ceiling, lighting the light is soft, the ingenious combination of light and shadow, the entire ceiling in one after another on the interpretation of the rhythm of the dance.

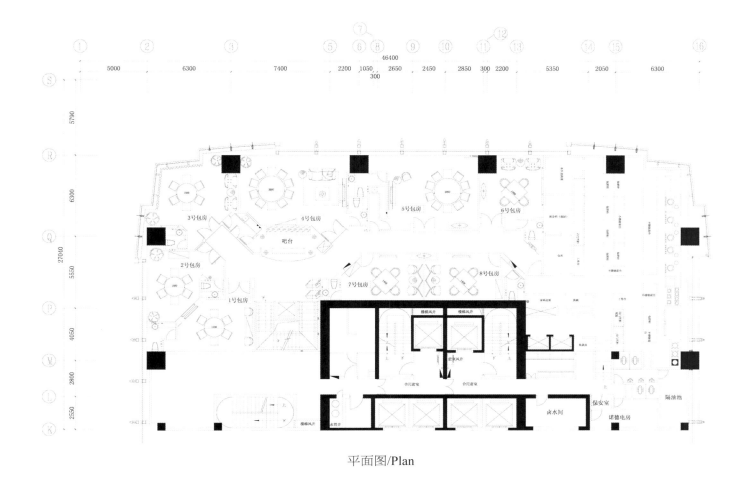

平面图/Plan

餐厅奢华而不艳俗,高雅而不冷艳;尽显高贵中的淡定从容,追寻高端的享受与飘逸。复古中的华丽,奢华中的风尚,雅致中的张狂,喧哗中的宁静。"金屋国菜"让客人在品尝美味佳肴的同时,亦享受贵族般的尊贵时尚生活。

Without the gaudy restaurants of luxury, elegance without ice; very calm and quiet and noble in search of high-end and enjoy the elegant. Gorgeous retro, luxury in the fashion, elegant in the insolent, noisy in the quiet. "Jin Wu China Food" where guests can taste the delicious delicacies while also still enjoy the respect patrician lifestyle.

## 贾 立:

室内设计师、产品设计师；

立和空间设计事务所设计总监；

北京原创设计推广协会副秘书长；

英格兰中部大学伯明翰艺术与设计学院，室内设计与信息技术专业硕士（获得2004 UCE MA 杰出奖）。

## 近期主要荣膺：

2010年，法兰克福Tendence展受邀参展设计师；

2009年，"记忆凳"家具被今日美术馆永久收藏；

2009年，"曲线"家具设计被中国国家大剧院永久收藏；

2009年，100% CRYSTALLIZEDTM 施华洛世奇水晶元素邀请合作设计师；

2009年，100% Design Shanghai DESIGN IN CHINA 设计大赛入围奖；

2009年，嫣然天使基金年度慈善拍卖小额义卖品邀请设计师；

2008-2009年度，第17/16届亚太区室内设计大奖APIDA提名；

2008年，第七届中国国际室内设计双年展CIID金奖；

2008年，"金外滩"最佳酒店设计提名奖；

2007年，"中国酒店设计大师赛"最佳概念设计奖。

# 荷畔餐厅

项目面积：*430m²*

项目地址：北京

竣工时间：*2010年8月*

设计公司：立和空间设计事务所

主案设计：贾立

摄影：高寒

隐于城市中的荷畔餐厅，就像它所在的位置一样，气质内敛、安静舒适。

荷畔位于北京西直门华远企业号院内。院子相对独立，四周被建筑围合。设计师通过室外木地板和新栽种的绿植，将餐厅入口向外延伸，同时将室外美景引入室内。餐厅室内空间本是不规则型，设计师利用这一特点，通过10米长的吧台将空间重新划分。吧台与操作台面相连，在墙面交接的位置，自然形成安静的水面。通过投影灯投射在原石墙面上的"荷畔"二字，若隐若现于水中，让外界的嘈杂瞬间无影无踪。

Hidden in the cities, LAKEVIEW Restaurants, the same as its location, is temperament, introverted, quiet and comfortable.
The relative independence of the restaurant's courtyard, surrounded by the building enclosed. Designers through the Outdoor wood floors and plants, will be an extension of the restaurant outside the entrance, while the introduction of indoor and outdoor views. The restaurant interior is irregular, designers use this feature, by redrawing the space bar. Bar is connected with the operating table, the location of the transfer in the wall, the natural formation of quiet water. Projected by the lamp surface in the original stone wall "LAKEVIEW" word, indeed, away from the noisy outside world.

距离餐厅不远处是一处安静的湖面，同时，由于"荷畔"的名字很容易让人联想到水和河岸的关系。水的涟漪成为室内设计的语言。不同颜色的木块通过拼接，形成餐厅的天花、墙面和地板。餐厅南侧是通透的玻璃幕墙，室外树木、土地与室内绿色和咖啡色相得益彰。北侧的咖啡区，临近墙面种满了喜阴的绿萝，通过镜面反射，让人恍若置身庭院中。

Restaurant next to a quiet lake. At the same time, "LAKEVIEW" name is reminiscent of the relationship between water and riparian. Then, the "water" into the language of interior design. By splicing different colors of wood to form a restaurant ceilings, walls and floors. The south side of the restaurant is the transparent glass walls, outdoor trees, green and brown land and interior complement each other. The north side of the coffee area, near the wall full of plants, through the mirror reflection, people think exposure to the courtyard.

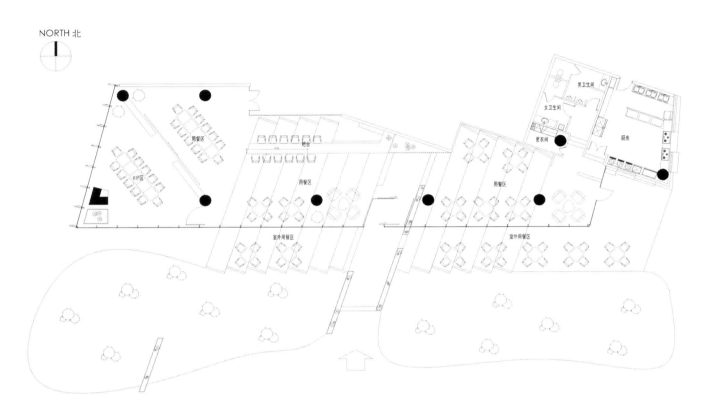

平面图/Plan

**周伟：**

周伟建筑工作室设计总监；

中国建筑学会室内设计分会杭州专业委员会理事。

**近期主要荣膺：**

APIDA第16届亚太室内设计大赛荣誉奖；

第六届《现代装饰》国际传媒大赛"年度商业空间大奖"；

首届"新设计，新环保"全国设计大赛室内设计组第一名；

中国50位最具商业价值的设计师。

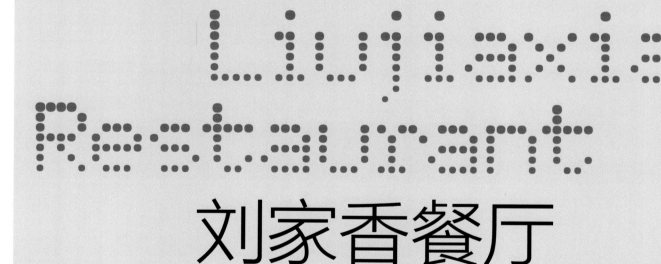

# Liujiaxiang Restaurant

# 刘家香餐厅

项目面积：*1200m²*
项目地点：中国杭州下沙大学城区
设计公司：周伟建筑工作室
主案设计：周伟
摄影：贾方

本餐厅的主要目标客户为大学城的老师和学生，设计师选择了苹果绿作为餐厅的主色调，给人一种青春阳光和希望的感觉，明确定位本案为一家年轻时尚的餐厅。1200平方米的餐厅分上下两层，在空间布局上和大多数餐厅一样，一楼为散座区，二楼为包厢区。进门楼梯下的不锈钢荷叶装置，以及一系列写实民工雕塑，为空间增添了不少的艺术气息。右边特别设计的小型宴会厅是专门为学生聚餐准备的，现代风格的家具和极具诗意的灯具让这个空间生动别致。

The main target customers of the restaurant are teachers and students of University. Designers chose the apple green as the restaurant's main colors, giving a feeling of youth and hope the sun. 1,200 square meters of the restaurant two storeys, in the space layout and most of the restaurants, like the first floor as a casual seat area on the second floor for the box area.Stainless steel door under the stairs lotus plant, and a series of realistic sculpture of migrant workers, for the space to add a lot of artistic temperament. The right side of the specially designed small banquet hall is designed to prepare dinner for the students, modern style furniture and highly poetic dramatic lighting and unique to this space.

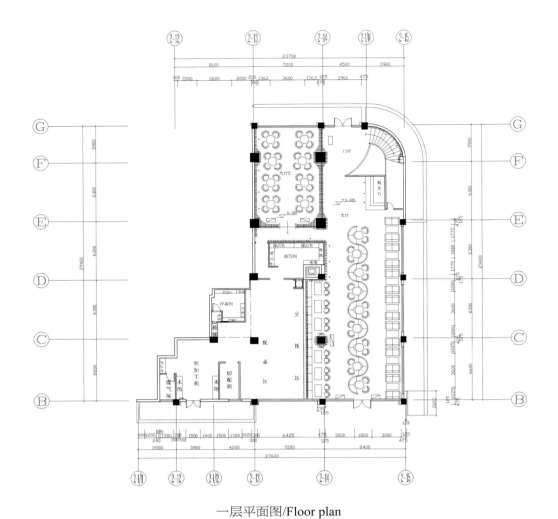

一层平面图/Floor plan

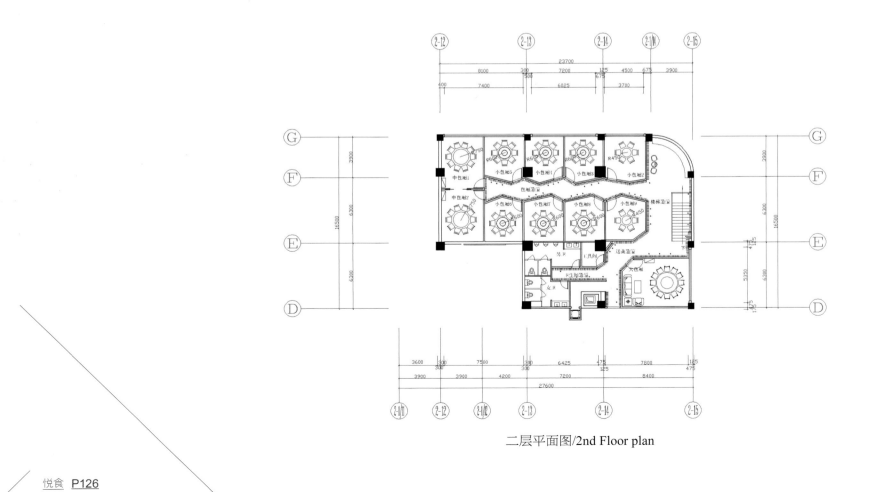

二层平面图/2nd Floor plan

大厅散座区顶面的设计手法一直延续到墙面，使整个空间焕然一体。大面积线条和镜面的运用为空间更加增添了时尚和迷离的感觉。三种不同的就座方式给客人提供了多种选择，s型就座区相互渗透，彼此观望，让客人就餐的同时也彼此互动。

Hall of the top surface area for odd design technique has been extended to the walls, large mirror lines and the use of space to add even more fashion sense. Took three different ways to give guests a wide selection, s-type seating area of mutual penetration, watching each other, let the guests eating at the same time interact with each other.

二楼走廊设计师通过平面动线的曲折和墙面的起伏,使空间的纵深感拉长,增加了空间的趣味性! 洗手间墙面同色调不同色相的烤漆玻璃的运用也是本案的亮点之一。

Plane moving through the second floor corridor designer line twists and ups and downs of the wall, stretched feeling of depth of space, increasing the interest of space! Bathroom wall paint with the colors of different hues of glass is the use of the characteristics of the case.

## 潘鸿彬：

潘鸿彬于香港理工大学设计学院分别取得室内设计学荣誉学士及设计学硕士，于2003年创立室内及品牌设计公司PANORAMA泛纳设计事务所。泛纳设计事务所创始人；香港理工大学设计学院助理教授；香港室内设计协会会长；香港设计中心董事。

## 近期主要荣膺：

美国IDA 设计大奖；iF中国设计大奖；中国最成功设计大奖；金指环-iC@ward全球室内设计大奖；APIDA亚太室内设计大奖，香港十大杰出设计师大奖；HKDA香港设计师协会设计双年奖；PDRA透视室内设计大赏；入围DFA亚洲最具影响力设计大奖。

## 谢健生：

谢健生于香港理工大学设计学院取得美术及设计学荣誉学士，香港室内设计协会专业会员。谢氏熟悉品牌设计、空间与管理的关系，在他的设计项目中充分地发挥及运用。谢氏担任香港生产力培训学院讲师，使室内设计师的角色得到更多社会人士的认识。

谢氏于2005加盟本地著名室内及品牌设计公司PANORAMA International Ltd.，参予亚太区的大型商业发展项目。公司的主要客户包括客户包括香港加州红集团、华润零售(集团)有限公司、中艺(香港)有限公司、香港Red Earth Production Ltd.、香港诗韵、香港海洋公园、香港新鸿基地产发展有限公司、深圳海港饮食管理集团、广东自由鸟服装有限公司、温州高邦集团等。

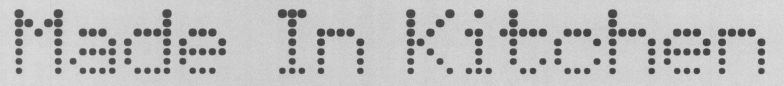

# 厨房制造

项目面积：*1116 m² (上下层)*
项目地址：徐州
设计组：潘鸿彬、谢健生、陈凯雯
客户：海港饮食策划管理有限公司
摄影：吴潇峰

"厨房制造"位于中国徐州繁华闹市,是一间新开设的中餐厅,为广大顾客供应各式精美新派中菜。

餐厅分上下两层,面积为1116平方米。 由于面向繁忙街道,使户外景观受到限制。 因此在设计中採用了独特的空间语言制造了各种人工景观,大大美化了餐厅内部环境,使顾客在享用美食佳肴时犹如置身于美丽的国画山水之中。

Located within the down town area in the city of Xuzhou, China. Made In Kitchen is a newly opened Chinese restaurant serving contemporary Chinese cuisine. With a total floor area of 1116 sqm. occupying two floors, the site imposed constraints of missing vertical linkage and unpleasant views of the busy street.

The design solution hence aimed at turning the spatial constraint into unique spatial languages and created the brand identity for this newly set up dining place. This was done by adopting design strategy of creating a total "inward looking" scheme with the help of a man-made landscape.

笼式梯井提供了与二楼的垂直连线, 打横的之字形图案围栏, 全高的黑木结构和隐藏的梯间灯槽营造出使人置身于国画中爬山登高的意境。 之字形落地白色金属屏风和随意点缀的方块形成了间隔和流动的感觉。 客人不同的视线可将自己与其他大众分开而享有充分的私人空间。

A minimal "cage like" stairwell was created to provide internal vertical linkage to the two levels. Full height dark timber structure with horizontal zic-zac pattern-like balustrade and concealed staircase light troughs suggested virtual experience of ascending a mountain in traditional Chinese painting. Zic-Zac arranged full-height white metal screens with random square patterns created screening effect and sense of "flow". Various diners' sight-lines and better privacy were provided for the 1/F open dining area.

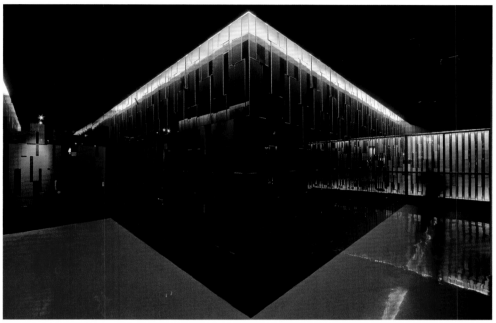

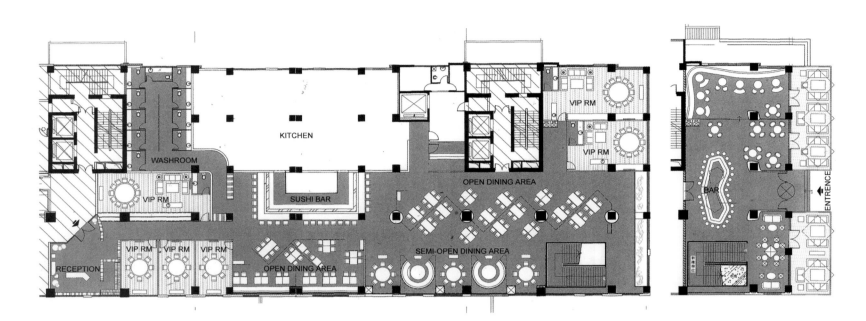

平面图/Floor plan

色调的改变演译了中国空间层次的定义，楼下是光亮的黄色而二楼是深红色。 私人空间分别由楼下敞开式的酒吧及二楼的公众大厅到半开放式的圆型包座而至私人贵宾房。

The Chinese definition of spatial hierarchy was established by the tonal change from bright yellow (G/F) to dark red (1/F) and level of privacy change from open (G/F's bar &1/F's open dining) to semi-open (circular booths) to private (VIP rooms).

## 潘鸿彬：

潘鸿彬于香港理工大学设计学院分别取得室内设计学荣誉学士及设计学硕士，于2003年创立室内及品牌设计公司PANORAMA泛纳设计事务所。泛纳设计事务所创始人；香港理工大学设计学院助理教授；香港室内设计协会会长；香港设计中心董事。

## 近期主要荣膺：

美国IDA 设计大奖；iF中国设计大奖；中国最成功设计大奖；金指环-iC@ward全球室内设计大奖；APIDA亚太室内设计大奖，香港十大杰出设计师大奖；HKDA香港设计师协会设计双年奖；PDRA透视室内设计大赏；入围DFA亚洲最具影响力设计大奖。

## 谢健生：

谢健生于香港理工大学设计学院取得美术及设计学荣誉学士，香港室内设计协会专业会员。谢氏熟悉品牌设计、空间与管理的关系，在他的设计项目中充分地发挥及运用。谢氏担任香港生产力培训学院讲师，使室内设计师的角色得到更多社会人士的认识。

谢氏于2005加盟本地著名室内及品牌设计公司PANORAMA International Ltd.，参予亚太区的大型商业发展项目。公司的主要客户包括客户包括香港加州红集团、华润零售(集团)有限公司、中艺(香港)有限公司、香港Red Earth Production Ltd.、香港诗韵、香港海洋公园、香港新鸿基地产发展有限公司、深圳海港饮食管理集团、广东自由鸟服装有限公司、温州高邦集团等。

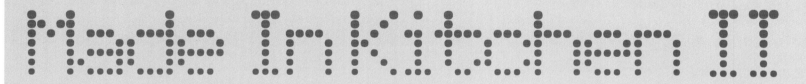

# 厨房制造II

项目面积：*4,000m²*
项目地址：中国芜湖
竣工时间：*2010年3月*
客户：海港饮食策划管理有限公司
设计组：潘鸿彬、谢健生、黄卓荣
摄影：吴瀟峰

厨房制造选址芜湖闹市，座落于市中心风景优美的湖畔，占地4000平方米，是此高档餐饮品牌最新开设的分店，为广大顾客提供当代中式美食佳肴。设计目标是把它打造成一所精致的美食餐厅，使嘉宾在享用佳肴美食的同时，可以从各不同方位感受美丽的湖光水色。大堂入口及走廊以波光水影，蝴蝶奇岩，水帘及浮莲组成精美的图案迎接嘉宾的光临。夹层楼面，圆型卡座及发光洒吧为10米高的中庭饮食空间提供不同的功能组合。主墙上的变色月亮在不同的时段营造出不同的情调，悬挂在空中的镜钢气泡及游鱼的投影令整个空间呈现出一幅悦目的影象。

Located within the down town area in the city of Wuhu, China, Made In Kitchen II is the newest roll-out of this high-end F&B brand serving contemporary Chinese cuisine. The site is facing a beautiful lake in the city centre with a total floor area of 4,000 sqm. The design strategy aimed at creating a unique dining experience by reinterpreting various beautiful scenes of a "lake". The resulting environment incorporated narrative elements in different zones. Motifs of "pool" + "ripples" + "butterflies" + "rocks" + "falling water" + "flying lotus" created a unique sense of arrival to the restaurant.Mezzanine floor, booth seating and central floating stage were introduced in the 10m high multi-functional atrium dining space to provide different sitting arrangements with different cosiness. The color-changing moon on the full-height feature wall provided different moods at different time. Floating mirror s/s "bubbles" and image projection of swimming fish multiplied the visual excitement of the whole space.

半高的玻璃间隔为顾客提供多种座位及半私人空间，中间的方型大枱及天花投影产生含蓄的视觉趣味及构成了整个空间的亮点。贵宾房内可以观赏一年四季的湖景，各种花草不同的形象和色彩表现了春夏秋冬的景色和情调，使顾客在享用美食时不再受到那种单调封闭的环境影响。

Low-height glazed partitions define various seating patterns and provide privacy to this open seating area. Ceiling image projection above central square table acted as focus & provided subtle visual interest to the area. The tonal changes of the lake in four seasons were represented in the private dining rooms. Images and colors of flowers representing spring, summer, autumn and winter provided different moods to the otherwise monotonous enclosed dining spaces.

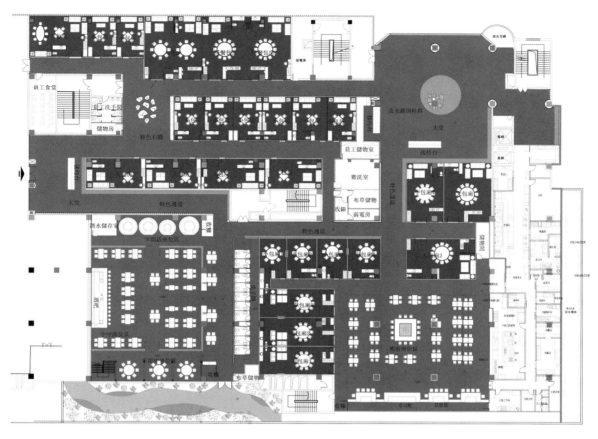

平面图/Plan

**曹 成:**

　　深圳深港建设及深圳汇博设计总监、高级室内设计师、高级工程师、国家一级专案经理、环境艺术设计师。

**近期主要荣膺:**

　　2005-2009年，广东省优秀建筑装饰工程奖（5项）。

　　2002-2009年，深圳市建筑装饰金鹏奖（10项）。

　　2006年，第二届中国（深圳）文博会首届中国创意设计大奖银奖（1项）和优秀奖（3项）；深圳市第七届装饰作品设计展一、二、叁等奖、佳作奖；2006第二届中国室内设计艺术观摩展最具创意设计奖；2006第二届中国室内设计艺术观摩展"方振华"最具创意设计大奖；2006第二届中国室内设计艺术观摩展室内设计十大新锐人物；2006年度APIDA亚太室内设计金奖及铜奖（香港室内设计协会）；"华耐杯"中国室内设计大奖赛优秀奖、佳作奖（中国建筑学会室内设计分会）；广东省优秀建筑装饰工程奖（2项）；2004至2006年度全国百名优秀室内建筑师（中国建筑学会室内设计分会）；2006年度精英设计师（现代装饰）。

　　2009年度杰出设计师大奖（现代装饰）。

　　2009年，第五届中国饭店业设计装饰大赛银奖。

# MR. PIGGER Restaurant
# 一口猪千禧大饭店

设计面积：*2600m²*
项目地点：北京东直门
主案设计：曹成

饭店原本1800平方米的空间由于净空的优势被加层扩建到2600平方米，这让业主欣喜若狂，因为这在北京东直门附近的高层楼宇中并不多见。作为该饭店菜系中地域体现的"猪"和"牡丹花"被用各种形式表现在静溢水雕塑、天花和立面墙体上，门厅及几个长廊的高低跌宕让空间的很自然的过渡，一定程度上做到了步移景迁。

Taking the advantage of the high headroom, the restaurant has been built up to 2600 sqm by adding mezzanine floors from the basic floor plate of 1800 sqm, this addition makes the owner greatly excited, as it is rare among high-rise buildings near Dongzhimen, Beijing. The "pigs" and "peonies" embodied in the domain of the cuisines of this Restaurant are envisaged in tranquil water sculptures, ceilings and facade walls in various forms. The well-distributed heights of lobbies and long corridors are naturally transitional in space, making reference transfer in steps and scenes to some extents.

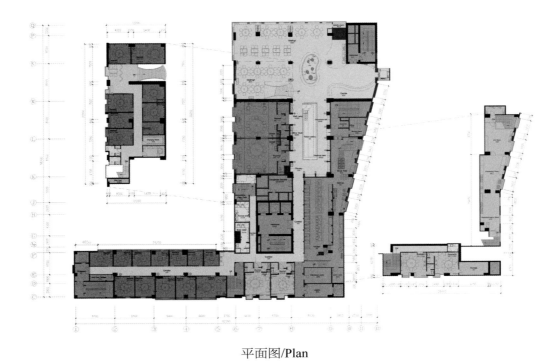

平面图/Plan

借鉴传统的欧洲风格的墙壁镶板来表现墙面，这种镶板通过丰富多彩的木材、铜制金属和仿古镜的采用而完善并逐渐现代化。

With wall panel boards in the traditional European style representing wall surfaces, this kind of panel boards can be perfected and gradually modernized using colorful timber, copper-metals and antique mirrors.

饭店的每一个区域都采用了对同一主题不同版本的设计表现，这种中西结合的风格让人觉得很熟悉。

The design expression in the same theme but of different versions is applied through out areas of the Restaurant. This style of the Chinese and Western combination is well-known.

除了镶板上的几何图形，具有特色的花卉图案（采用标志里的牡丹花图案）也在整个空间中得到运用，它通过不同颜色被用于雕塑、地毯、门把手的细节部分以及楼梯扶手和屏风中。

In addition to geometric pattern on panel boards, the characteristic of flower patterns (using patterns of peonies for the marks) are applied in the space as a whole, and are used in detailed parts of sculptures, carpets and door handles as well as in staircase handrails and screens in different colors.

**陈彬：**

武汉理工大学艺术设计学院副教授

中国建筑学会室内设计分会（CIID）会员

中国建筑装饰协会会员及设计委员会委员

国际室内建筑师/设计师联盟（IFI）专业会员

亚太建筑师与室内设计师联盟（IAI）理事

大木（湖北）后象设计顾问机构 设计主持

## 近期主要荣膺：

作品入选ANDREW MARTIN 2010年度国际室内设计大奖；

2010年，亚太室内设计双年大奖赛-评审团特别大奖；

2010年度，广州国际设计周"金堂奖"年度十佳餐饮空间设计；

2009年度，APIDA第十七届亚太区室内设计大奖银奖；

2009年度，作品入选德国iF2009中国设计大奖；

2009年度，广州国际设计周"金羊奖"中国十大室内设计师；

2009年度，广州国际设计周"金堂奖"餐饮酒吧类中国十大设计师；

2009年度，Interior Design China "酒店餐厅类最佳设计"奖；

2009年度，金指环全球室内设计大赛优秀奖。

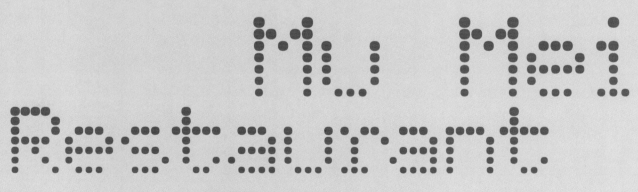

# 慕美餐坊

项目面积：*1030m²*

项目地点：中国深圳

完成设计时间：*2010年1月*

设计公司：大木设计中国(湖北)后象设计顾问机构

主持设计师：陈彬

参与设计师：李健、傅晟、李丹

摄影：吴辉

餐厅在南方的深圳，店名很妙。

设计师选择了温润淡雅的色调来营造空间氛围。绿宝和雅士白云石使公共空间显的冰清玉洁；浅色橡木和灰绿地毯的组合，又让就餐区亲和而宁静；尺度非常的灯饰优雅迷人，在视觉上传达着时装的魅力。整个餐厅流动着一种特殊的女性气质，正如她的名字——慕美。

Situated on the south of China in Shenzhen, the art dining hall has a wonderful and spiritual name.

Designers take warm and quiet color as the main tone of the whole restaurant. Verde jade and airston marble are used to express the pureness and elegance of public area, in the similar way, the combinations of light oak and celadon carpet make the dining place snug and tranquil, and exaggerated light-fixtures convey the glamour of fashion visually. A particular feminity flows along the whole space, to which the owner has endowed it the name of "Mumei".

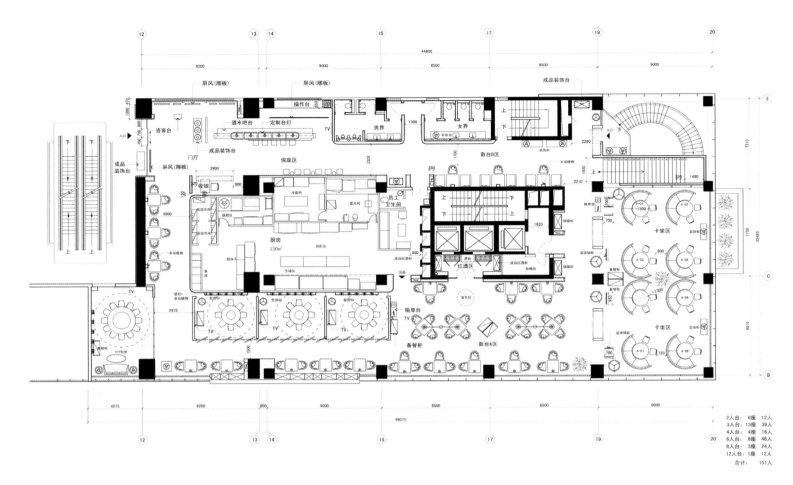

一层平面图/Floor Plan

| | | |
|---|---|---|
| 2人台: | 6座 | 12人 |
| 3人台: | 13座 | 39人 |
| 4人台: | 4座 | 16人 |
| 6人台: | 8座 | 48人 |
| 8人台: | 3座 | 24人 |
| 12人台: | 1座 | 12人 |
| 合计: | | 151人 |

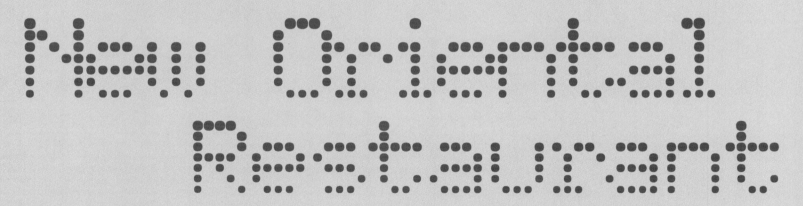

# New Oriental Restaurant

# 新东方火锅店

项目面积：*1200m²*

项目地址：长春市和顺街4号

主创设计：李文

主要材料：青花瓷瓶、灰瓦、藤条、水泥、镜片

摄影：李晓军

本方案以中国传统文化中的水墨丹青和如意瓶造型元素为基本设计理念，结合现代表现形式，充分体现材料与现代空间的结合。通过对白色天棚和地面（宣纸）灰瓦、水泥（墨）、红色藤条（丹）以及青花瓷盘（青）的运用组合，力求营造出一个充满中国水墨画境的概念性新餐饮环境。

The program in traditional Chinese culture bottles of ink and the wishful concept of modeling the basic design elements, combined with modern forms, to fully reflect the combination of materials and modern space. On use of the white ceiling and floor and gray tiles, cement, red cane and blue and white porcelain plate combination, and strive to create an environment filled with Chinese ink painting new conceptual dining environment.

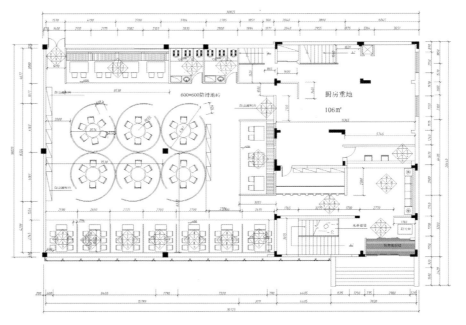

一层平面图/Floor plan

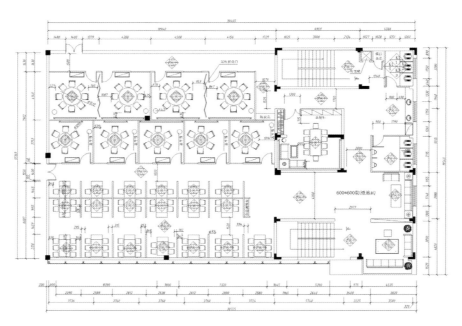

二层平面图/2nd floor plan

## 赖建安：

生于中国台湾，国立中兴大学土木工程专业。

台湾何肇熹建筑师事务所；台湾安之居建筑师事务所。

1998年，杭州凯悦专案设计师；

1999-2000年，受邀规划设计上海龙通生活广场；

2001年，受邀规划设计-K.K.Marriott.万豪大酒店（Malaysia）；

2002-至今，十方圆国际设计工程公司担任设计总监；

2007年，受邀施工图扩出及施工-上海空间美学馆（Cassina）；

2007年，受邀规划设计-上海Iswin Park-5A Office大楼及林氏博物馆会所；

2008年，受邀规划设计及施工-"我型我秀"制作单位New Office（上海上腾娱乐有限公司）；

2009年，受邀规划设计-Yeng Hotel五星级大酒店（Malaysia）；

2009年，受邀规划设计-Taiwan赖敏树博物馆会所；

2008年-至今，集慧堂装置艺术设计工程公司担任总监。

## 近期主要荣膺：

2005年，入选"2005名家室内设计选/当代杂志"；

2008年，"2008台湾室内设计大奖得奖作品选/TID"；

2008年，作品入选"2008中国环境设计年鉴"。

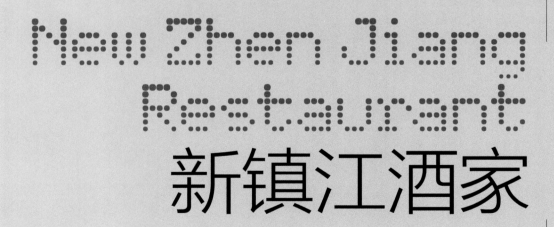

新镇江酒家

项目地址：南京西路1111号
设计人员：赖建安、高天金
设计单位：十方圆国际设计工程
施工单位：十方圆国际设计工程

品——赏食空间。本案地处上海市南京西路——沪上顶级精品百货区段。规划面积仅100平方米,是家老字号上海本帮菜。有着经营餐饮业数十年经验的业主,极力支持成为新镇江新风貌,以私人会所形态演绎后现代中国风格呈现。

牡丹花里······"惟有牡丹真国色,开花时节动京城。"全案以中国民间最具代表性的牡丹花作为亮点,增添了该案的东方人文色彩。花代表着花开富贵之吉祥寓意,呈现后现代中国风之风格。

Taste——Enjoy the food space。Case, fine department stores located in Shanghai's top section, planning area of only 100 sqm. Shanghai Restaurant is a main food shop. Business owners have decades of experience in catering, restaurants, private clubs form of interpretation of the present post-modern Chinese style.
Peony in······"The peony is Aromatic, when she bloom, it will disturb the capital." Peony flower representing China, adding to oriental human color. Blossoming flower represents the mean.

空间······
入口透着穿透式的圆形拱门拉开空间序幕，透光玻璃酒架形成了包房的私密空间。在东方底蕴包涵下设计师为求精致、现代感，利用金属镀钛球、S.T.亮片，及灯光的背景衬托，使得桌上大理石及餐具与主食成为主角，更亲自挥洒墙面艺术及极富空间纵深的肌理油画。

Space······
Reveals the entrance arches, glass wine racks formed a private room private space. Bear with a designer in the East heritage for the sake of fine, contemporary, using metal titanium ball, ST sequins, and lighting the background, making the marble tableware become the protagonist.

淬炼……
设计师采用后现代主义的精神及技术，寻找新的设计语言，大胆提取受人喜爱的奢华精神与文化，诠释出后现代主义的精髓。每个物件的线条中、每个饰品的理念间、在每件装置的背后、所有细节里闪耀着设计师的用心。

Temper……
Designers using the spirit of post-modernism and technology, looking for new design language, extracted by the bold spirit and luxury of popular culture, and interpret the essence of postmodernism. The lines in each object, the concept of jewelry in each room, the back of each device, shining all the details of the designer's intentions.

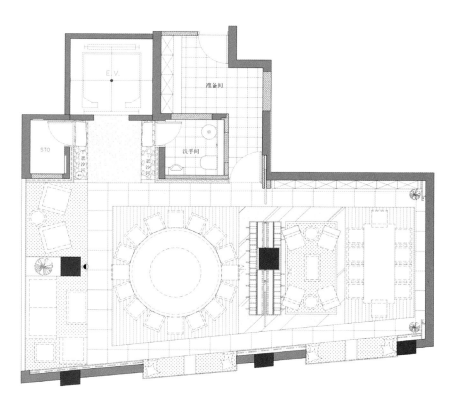

平面图/Plan

**陈彬：**

武汉理工大学艺术设计学院副教授

中国建筑学会室内设计分会（CIID）会员

中国建筑装饰协会会员及设计委员会委员

国际室内建筑师/设计师联盟（IFI）专业会员

亚太建筑师与室内设计师联盟（IAI）理事

大木（湖北）后象设计顾问机构 设计主持

**近期主要荣膺：**

作品入选ANDREW MARTIN 2010年度国际室内设计大奖；

2010年，亚太室内设计双年大奖赛-评审团特别大奖；

2010年度，广州国际设计周"金堂奖"年度十佳餐饮空间设计；

2009年度，APIDA第十七届亚太区室内设计大奖银奖；

2009年度，作品入选德国iF2009中国设计大奖；

2009年度，广州国际设计周"金羊奖"中国十大室内设计师；

2009年度，广州国际设计周"金堂奖"餐饮酒吧类中国十大设计师；

2009年度，Interior Design China "酒店餐厅类最佳设计"奖；

2009年度，金指环全球室内设计大赛优秀奖。

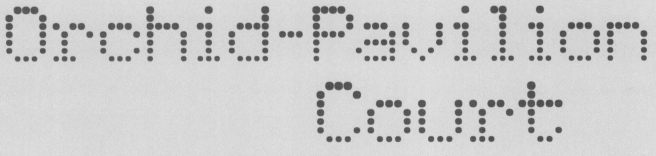

# 兰亭别院

项目面积：660m²

项目地址：中国武汉

完成设计时间：2009年5月

设计公司：大木设计中国(湖北)后象设计顾问机构

设计师：陈彬

参与设计：李健、傅晟、李丹

摄影：吴辉

兰亭别院是一间供应中式餐饮和茶点的复式餐厅，位于武汉汉口江滩。在这个空间中，设计师求取西晋王羲之《兰亭序》中"曲水流觞，茂林修竹"的意境，运用贴近自然的材料和平实的手法，创造一个追求高古情怀的写意空间。在满足食客对美食要求之外，更能带给其视觉和精神的享受。

Located on the River Bund of Hankou, Wuhan, *Orchid-Pavilion Court* is a duplex restaurant serving Chinese food and refreshments. In its space, designers introduce the ideal surroundings of *cup floating down the curving stream, trees with thick foliage and tall bamboos* as described in *The Orchid Pavilion by* Wang Xizhi, a famous Chinese calligrapher in Western Jin Dynasty and employ the almost natural materials and plain crafts to create a free and relaxing space pursuing for elegance, simplicity and ancientness. Customers are served with satisfactory cuisines, more significantly, along with visual and spiritual delight.

餐厅面积666平方米，设计师将其划分为三个不同功能区域：一、接待区；二、明档及取食区；三、进餐区，而进餐区又规划为三种形式，即散座区、卡包区和包房区，以满足不同食客的需求。为了回避独层空间的单调感，设计师首先利用两排丝印玻璃围合的卡包，将整体空间分割开来，使视线受到相应阻碍，同时墨竹丝印玻璃又成为整个餐厅的视觉主题，掩映出"茂林修竹"的情境氛围；其次又将餐厅中部区域抬升地坪，形成抬高区，上置高背沙发，四周以黑色云石为水面，以断面石料为山石，运用现代构成理念铺设，演绎出"曲水流觞"的高雅情趣，抛其简单形似，追求内在神似。

Covering an area of 666sqm, *Orchid-Pavilion Court* is divided into three different functionary areas: reception area, food-service area and dining areas. To meet the needs of various customers, Dining area is composed of free-seat sector, booth sector and private rooms. To avoid the monotony of one-story space, designers use two rows of silk-screen glass to enclose the booth seats, segregate the whole space, and produce corresponding sight obstacles firstly, bamboo silk-screen glass becoming the visual theme of the whole restaurant and reflecting *the surroundings of thick foliage and tall bamboos*. More importantly, such design uplifts the central area of the restaurant to form elevated areas, on which high-back sofas are placed, surrounded by black marble as water surface and cross-section stones as rockery. Employing modern constructivism notion, it fully interprets the elegant delight of *cup floating down the curving stream*. Right here, the simple shape likeness gives way to inner spiritual likeness.

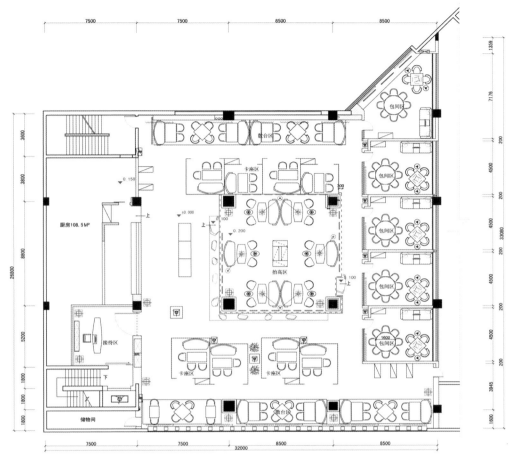

平面图/Plan

设计师特别选配了龙泉青瓷为兰亭别院的瓷具用品，并将整个餐厅空间以温暖原木色系为主要调性，不同纹理和深浅的木面运用于不同的装饰空间中，天面、墙面、地面和家具，利用厚重的暖调去映衬碧绿的龙泉青瓷，使原木的温厚和青瓷的清纯，在空间中相得益彰。此外，粗面亚麻布面也运用在灯体、隔帘、墙面、吊顶和家具面料等不同的地方，亚麻布亲和纯朴的气质，拉近了食客与空间的距离，特别的手感又再次映衬出青瓷玉石般的细腻触感，给使用者视觉和触觉的双重享受。另外，布置于餐厅中符合高古审美趣味的粗陶和石雕物件，不时静静点击食客视线。上古雅趣，玄风禅意弥满在美食美茶的空间里。

Designers particularly select Longquan blue porcelain as the porcelain supplies and take warm wood color as the main tone of the whole restaurant. And different textures and shades of wood surface are used in various decoration spaces, such as ceiling, walls, floors and furniture. The heavy warm color tunes against the transparent Longquan blue porcelain makes the generousness of wood complements very well with the ignorance of blue porcelain in this space. In addition, the rough linen cloth is also used in various places such as lamp-body, separating curtains, walls, ceiling, and furniture fabrics. Computability and simplicity of linen cloth narrowed the distance between diners and space. Special touch once again reflects a delicate jade-like touch of blue porcelain, giving users dual visual and tactile enjoyment. What's more, rough ancient pottery and stone-craving objects of simple and elegant aesthetic taste arranged in the restaurant occasionally catch the customers' attention. While enjoying first-class cuisine, customers may taste ancient elegance and possibly gain some wisdom from Chinese Taoism and Buddhism as well

**张奇永：**

1982年，出生于黑龙江省哈尔滨市；

1999年，就读于黑龙江建筑职业技术学院；

2002年6月—11月，就职于中央美术学院工作二室；

2002年11月—2003年5月，就职于上海市金陵东路某家装公司；

2003年5月，就职于哈尔滨唯美源；

2005年5月，唯美源设计兼摄影至今。

**近期主要作品：**

2005年5月，现代粗粮（大连）；

2005年12月，锦州大连（锦州）；

2006年10月，现代食尚（大连）；

2006年，人和（哈尔滨）；

2007年，女王（哈尔滨）。

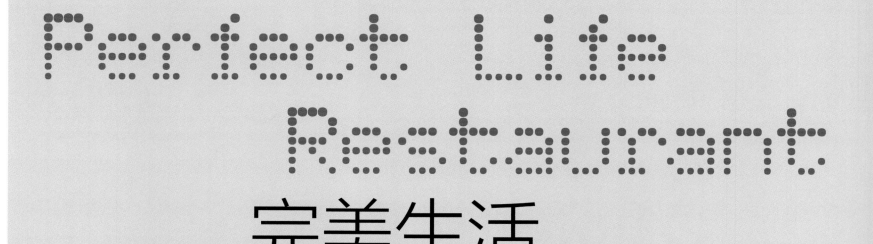

# 完美生活

项目面积：*600m²*
项目地址：齐齐哈尔龙沙路
设计师：张奇永
主要材料：石材、瓷砖、玻璃、地板

作为一个设计师，从来没有真正的审视过生活，忙忙碌碌的一天只认为设计是生活重要的一部分，有的时候与其说是生活不如说是活着。忙着去搞方案、忙着去实现自己设计的思想。嘴里面经常谈着的是设计离不开生活，但何时真正的去观察过生活，审视过生活。今年偶然的一个机会，接下了一个餐厅设计。这个餐厅的名字起得有些感性，和餐厅配合起来有些不搭调，叫"完美生活"。做着设计之余也想着生活问题，生活离不开乐观的面对每一天，而每一天又在品味着生活，感悟着设计。但是说突然有一天来了个设计叫完美生活，从创意上来讲不可能说把完美生活去怎样创意，而是说更多的通过这个案子去给自己一定的反思。

As a designer, had never really look at life, busy day. Sometimes life is not so much as it is alive. Busy to engage in programs designed for busy to realize their ideas. Mouth and face often talked of the design is inseparable from life, but had to observe when the real life, look at my life.A chance coincidence of this year, took over a restaurant design. The restaurant's name was some emotional, and restaurants together with some dissonant, called "perfect life." I also think the design of life issues, can not live without optimism in the face every day and every day and in the taste of life and feeling design. Suddenly one day come to a design called "perfect life", from the creative speaking to can not say what the perfect creative life, but rather that more of their own through the case to go to a certain degree of reflection.

完美生活是具有空间想象的，当中的"完美生活"是合乎人性的，合乎于生命的东西，但是完美的东西都是相对的而不是绝对的。作为一个小空间的餐馆，能提倡这种理念的设计，非常值得我对他重新认识！我尽我所能，来体现我所想象完美生活，想它简洁的四壁，每天都有充足的阳光透过宽敞的大玻璃洒向室内，让走进来的人对生活有着无限想象的完美生活。

Life is a perfect space for imagination, which the "perfect life" is in line with human nature, in line with the things in life, but perfect things are relative, not absolute. As a small space, restaurants, to promote the design of this concept is very worthy of my new understanding of him! I do my best to reflect my idea of a perfect life, want it simple walls, plenty of sunshine every day through the large glass beams and spacious interior, so that people coming into the limitless imagination of life perfect life.

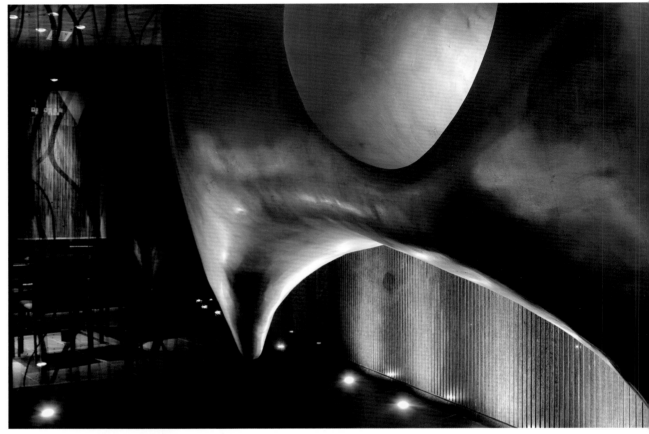

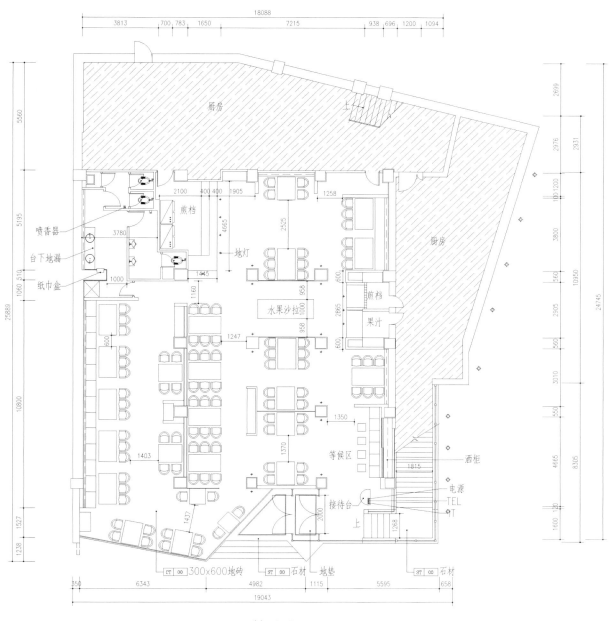

一楼平面图/Floor plan

生活离不开灿烂的阳光，自己认为天天熬夜作案子的设计师，特别渴望着能有闲暇的时间，可以真正的观察生活。原来说的自认为对生活很有观察，其实并没有特别静的去看待生活！看看那一片片的树叶，听听森林里那自然的声音，感受自然的气息，这是大家都向往的。当真正有一天休息的时候，我要去追求这种完美生活，我认为完美生活是永远追求不到的，完美生活只是精神层面的。但是我觉得大家在忙碌之余，起码要想到还有对完美生活的一种向往就够了。

Can not live without sunshine, staying up late every day, son of the designer crime, especially eager to have leisure time, to observe the real life. Had said that life is very self-observation, in fact, not particularly quiet to look at life! To see which pieces of leaves and listen to the voice of the forest that nature, feel the natural atmosphere, this is all long for. When the real day of rest, I'm going to pursue this perfect life, I think that the pursuit of the perfect life is not forever, the perfect life is spiritual. But I think that we more than in the busy, at least to think there is a yearning for a perfect life is enough.

完美的生活是享受着自然。自然界给人创造了阳光、空气、水，包括食物。如果这些物质生活大家可以用自己良好的精神状态把它体现到生活的更高层面，我想这就是完美生活！

A perfect life is enjoying the nature. Nature gives to create the sun, air, water, including food. If the material you can use your own good mental state to the life it reflects a higher level, I think this is the perfect life!

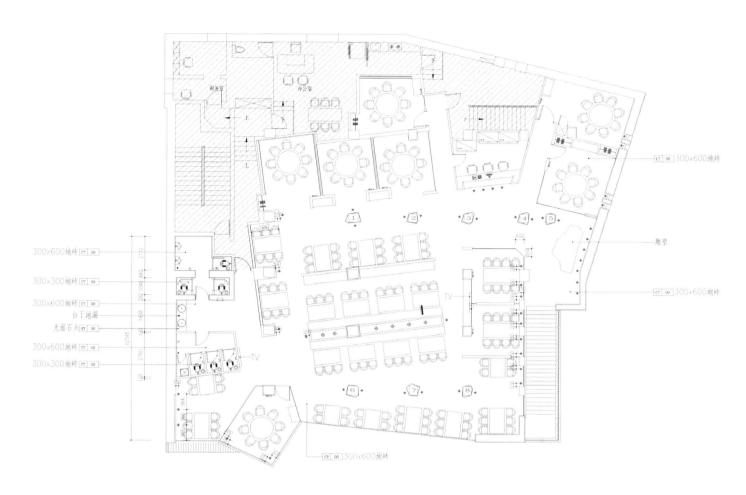

二楼平面图/2nd floor plan

**方令加：**

三佰舍室内设计设计总监

1981年-1993年　农民

1993年-1997年　工人

1997年-2000年　学徒

2000年-2003年　无业

2003年-2007年　喜马拉雅设计师

2007年组建三佰舍室内设计顾问有限公司

# Qingtang
# Restaurant

# 清汤餐厅

项目面积：**480m²**

项目地址：福建　厦门市

竣工时间：**2010年9月**

设计师：方令加

参与设计师：李少东

设计单位：三佰舍室内设计顾问有限公司

餐厅的环境，和料理一样，清淡、朴素，是现代的也是中国的。

大空间的关系根据功能需求分隔为不同大小的各个方形空间，型体上并无做多余的修饰，以使整体框架是纯粹和现代的，墙角的黑边以使墙面不会被大面的白色溶化，保护了墙角也使空间的骨架上带有最为简洁的中国味道。把对比强烈的或老或新的家具及摆件置入空间，使传统和现代在干净的空间中交错游走。

Restaurant environment, like food, light and simple, is also a modern and also is China.

The relationship between the large space separated according to functional requirements for different sizes of each square space, type of body modification is not made redundant, so that the overall framework is a pure and modern, black side wall so the wall will not be large areas of White melt and protect the wall to make room for the skeleton is also the most compact with Chinese flavor.The contrast or old or new furniture and ornaments into space, traditional and modern in a clean space staggered walk.

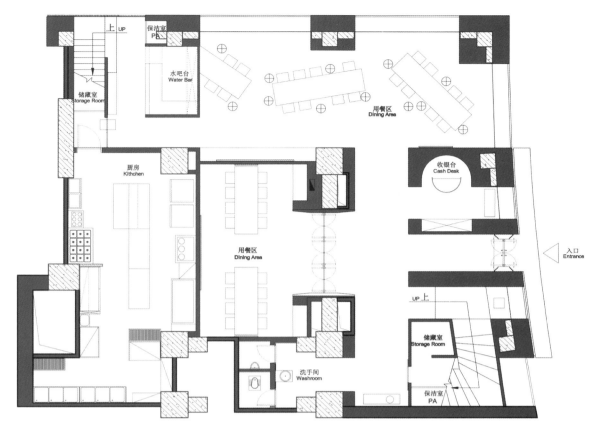

一层平面图/Floor Plan

二层平面图/2nd Floor Plan

# SCENA意大利餐厅

项目名称：*SCENA意大利餐厅*
项目地址：上海浦东丽思卡尔顿酒店52层
设计公司：*Super Potato*
竣工时间：*2010年6月*
图片提供：上海浦东丽思卡尔顿酒店

SCENA位于酒店的第52层，是由日本著名设计师公司Super Potato精心设计的意大利餐厅。餐厅的部分天花板和墙壁采用随意中空设计，让客人置身于一种幸福、亲和，细致的用餐氛围中，留连忘返。而餐厅的灯光效果更是一大创新，其内藏在天花及墙壁中的15部投影仪将餐厅气氛塑造得罗曼蒂克，温馨动人。

SCENA located on the 52nd floor, is a Italian restaurant designed by a famous Japanese designer Super Potato. Part of the ceiling and walls of the restaurant with casual hollow design, allowing guests exposure to a happy, pro-dining atmosphere and detail, will never forget. The restaurant's lighting is a major innovation, its built-in ceiling and walls of the 15 projectors will be shaping the romantic atmosphere of the restaurant and warm.

SCENA提供城中最佳的意大利轻松美食，全日供应各款新鲜菜肴、套餐、自助早餐和其他进餐时段的自助餐。SCENA主要用餐区域能容纳128位宾客。如需要更多私密空间，餐厅可提供16个座位的豪华贵宾房。

SCENA provide the best Italian food in town, all day sections of fresh food, set menus, buffet breakfast and other meal time buffet. SCENA the main dining area can accommodate 128 guests. If you need more private space, the restaurant offers 16-seat luxury VIP rooms.

朱君：

毕业于哈尔滨师范大学油画系

北京当代艺术画家

现任哈尔滨唯美源装饰设计有限公司设计师

# Shantuyuan Restaurant

# 膳福源

项目面积：*541m²*

项目地址：哈尔滨道里区安隆街

竣工时间：*2010年5月*

设计师：朱君

主要材料：光面黑色石材、喷砂玻璃、金色镜子、青玻璃、黑色不锈钢、不锈钢雕花、有色涂料、木质喷漆、人造皮革包框

本店位于繁华地带，经营者力求打造良好的消费环境，让顾客有低消费高品质的感觉。本店用不完整的美展现完整的美，例如用不完整的几何三角形为主题，无论是本店的外墙面造型还是里面的椅子，还有用三角形和镜子相结合展现出一种宽阔的视觉错误，让人们觉得本店的空间趣味性增强，这样大大的增加了本店的视觉空间，同时也让人们感受一种另类的美。

Restaurant located in the downtown area, operators strive to build good customer spending environment for customers with low consumption of high-quality feel. Shop with incomplete to show the full beauty of the United States, for example: the geometric triangle with an incomplete theme, whether it is the outer shape of the walls or restaurant chairs, a mirror with a combination of triangular and show a broad visual error, greatly increase the shop's visual space, but also allow people to feel a different type of beauty.

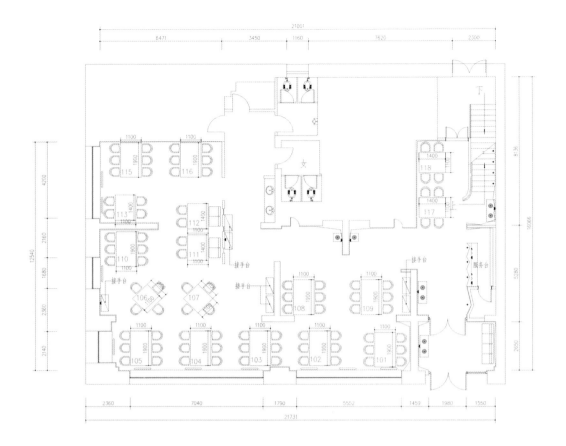

一层平面图/Floor plan

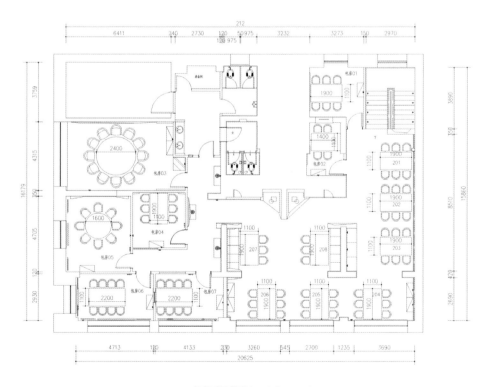

二层平面图/2nd floor plan

**陈彬：**

武汉理工大学艺术设计学院副教授

中国建筑学会室内设计分会（CIID）会员

中国建筑装饰协会会员及设计委员会委员

国际室内建筑师/设计师联盟（IFI）专业会员

亚太建筑师与室内设计师联盟（IAI）理事

大木（湖北）后象设计顾问机构 设计主持

## 近期主要荣膺：

作品入选ANDREW MARTIN 2010年度国际
室内设计大奖；

2010年，亚太室内设计双年大奖赛-评审团
特别大奖；

2010年度，广州国际设计周"金堂奖"年
度十佳餐饮空间设计；

2009年度，APIDA第十七届亚太区室内设计
大奖银奖；

2009年度，作品入选德国iF2009中国设计
大奖；

2009年度，广州国际设计周"金羊奖"中国
十大室内设计师；

2009年度，广州国际设计周"金堂奖"餐饮
酒吧类中国十大设计师；

2009年度，Interior Design China "酒店餐厅
类最佳设计"奖；

2009年度，金指环全球室内设计大赛优秀
奖。

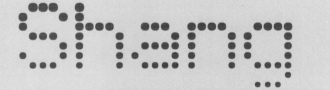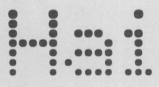

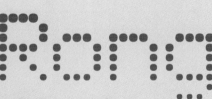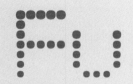

# 上海荣府

项目面积：*631m²*

项目地址：上海威海路

完成设计时间：*2009年10月*

设计公司：大木设计中国*(湖北)*后象设计顾问机构

主持设计师：陈彬

参与设计师：李健、傅晟、严小兵

摄影：吴辉

上海威海路，一栋保存完整的小洋房。设计的趣味在于如何在这个空间里营造一个契合当代商务精英新需求的美食空间，同时又能感受新旧上海的独特文化风尚。

Located on the road of Weihai in Shanghai, Rong Mansion is a well-preserved aged building. A lot of fun of the design lies in creating a unqiue space for elite business people, who are satisfied with new needs of cuisines,more significantly, along with the unique essence of both old and new Shanghai.

整个空间营造了许多的"冲突"：现代色与形的壁纸与
手工实木墙裙、丝印玻璃时尚花格与巴洛克风格的精
美木柱、透光膜天花与古典优雅气质吊灯、现代简约的
办公家私与雍容华贵的古典家具、纯净光洁的人造石
与呈现天然美感的大理石，时尚与古典之美的碰撞与
融合，让视线在时光中穿行，触碰久远的记忆，感知文
化的变迁。

Conflicts spread the whole space of the art
dining hall, in which modern wallpaper combined
with manual solid paneling, silk-screen glass of
burberry plaid go with Baroque delicate pole,
light-through ceiling mixed with elegant and
classical pendant lamp, modern office furniture
confronted with traditional aged ones, silky pure
white artificail stone complemented with marbel of
natural sense of beauty. Rong Masion brings you a
great interpretation of conflict and merge between
fashion and classic, where the vision wanders
through the course of time and remind us faded
memory and transmit of our culture.

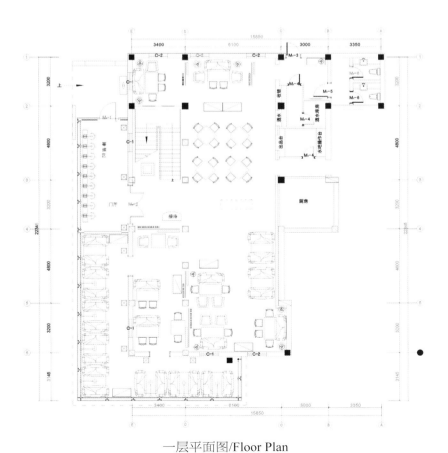

一层平面图/Floor Plan

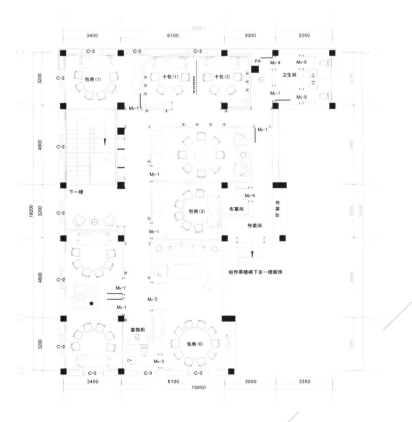

二层平面图/2nd Floor Plan

空间的视觉在光影之中穿梭，感受时空的逆转，触碰到那些久已褪色的记忆。

Among the visual space shuttle in the light, feel the space and time reversal, and touch to those who have long faded memory.

### 徐晓华:

中国建筑学会室内设计分会会员，国家注册室内建筑师。

毕业于苏州工艺美院,苏州徐晓华室内设计有限公司设计总监,擅长于酒店、餐饮、娱乐、休闲等大型娱乐场所的室内设计。

主设计项目有：苏州园外楼酒店、泰州嘉銮国际大酒店、苏州望湖宾馆、苏州美倫MEILUN会所、苏州一代娱乐城、苏州金色上海滩休闲会馆、苏州嘉年华休闲会馆、泰州鹏欣丽晶温泉会所、无锡红豆股份公司、芜湖鸠江区法院、芜湖鸠江人民检察院、张家港港务集团。

### 近期主要荣膺：

"苏州园外楼饭店"，荣获"2008年中国室内室内设计大奖赛"酒店、宾馆工程类二等奖、荣获"2008 INTERIOR DESIGN China 酒店设计奖"入围奖、荣获"16届亚太室内设计大奖"十名入围奖 。

"苏州徐晓华室内设计有限公司"，荣获2008年中国室内室内设计大奖赛办公工程类三等奖。

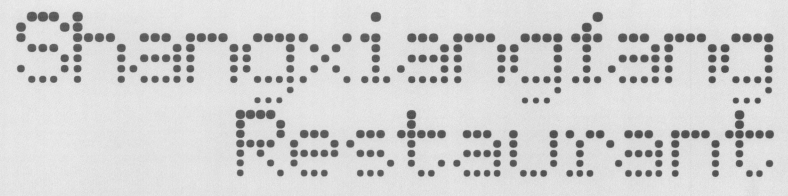

# 上厢房餐厅

项目面积：*1200m²*
项目地址：扬州广陵区跃进桥壹号广场首幢
主案设计：徐晓华
设计公司：苏州徐晓华室内设计有限公司
主要材料：饰面板、石材、瓷砖、壁纸
摄影：潘宇峰

如何突破传统的湘菜馆零乱中式风格,融入现代的设计手法,寻找企业文化与建筑空间的切合点,在传统中餐文化中寻找新的元素,在本案设计过程中进行了大胆的尝试。

How to break through the traditional Chinese style Hunan Restaurant disorder, into a modern design techniques to find the corporate culture and meet the point of architectural space, in the traditional Chinese food culture in the search for new elements, in this case the design process was a bold attempt.

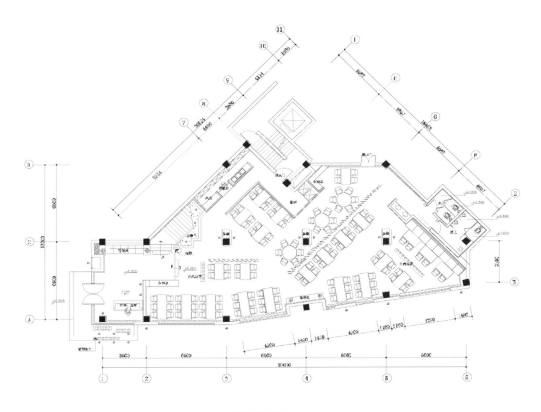

一层平面图/Floor plan

餐厅一楼为散座厅，二楼为包厢区。一楼散座厅，墙面用了大量的实木板，在顶面和隔断上也用了相同的材料，顶面造型的实木板从上至下巧妙分隔了空间，使餐位与餐位之间自然的分隔，形成了一个个相对私密的空间，虚与实，空与满在空间中处理的恰到好处。二楼包厢走廊，用了白橡饰面，使得餐厅的色调更加统一、温馨。

Block for the casual restaurant on the first floor hall, second floor balcony area. First floor powder room blocks, wall panels with a lot of wood, cut off the top and also used the same material, the top surface shape of the wood panels separate the space under the clever, the seating and the natural separation between seats, a relatively intimate form a space, virtual and real, empty and full processing in the space just right. Box in the gallery corridor, with a white oak finish, making the colors more uniform and warm.

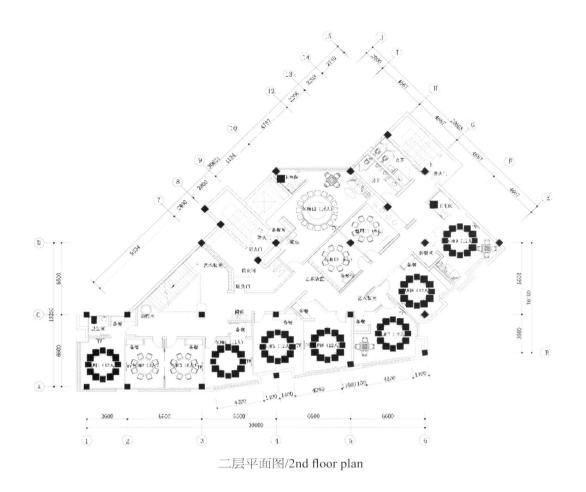

二层平面图/2nd floor plan

## 利旭恒：

出生于中国台湾，英国伦敦艺术大学 BA (Hons) 荣誉学士。

GOLUCCI DESIGN LIMUTED古鲁奇公司设计总监；长年致力于酒店餐饮空间、专卖店与地产商的设计工作，10多年的酒店餐饮零售空间设计，累积了丰富的项目经验，在多个设计风格与商业形态上都卓有成就。

## 近期主要荣膺：

2009年度金外滩最佳餐厨空间优秀设计奖（美国室内Interior Design）；

2008年度中国十大样板间设计师（中国建筑装饰协会）；十佳配饰设计师（中国第二届）；

2008年，作品鼎鼎香餐厅获北京最佳火锅餐厅称号（City Weekend）；

2007年，作品鼎鼎香餐厅获北京最佳餐厅称号（That's Beijing）；

2006年度中国十佳设计师（中国建筑装饰协会）；

2004年度中国十佳设计师（中国住交会）。

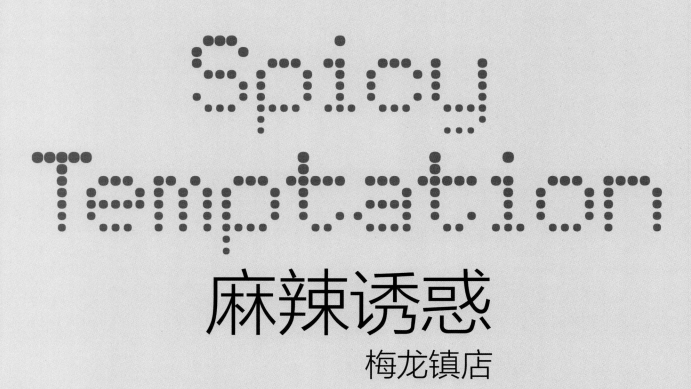

# Spicy Temptation

# 麻辣诱惑

## 梅龙镇店

项目面积：*800 m²*

项目地址：上海梅龙镇广场*7*层

主创设计：利旭恒

参与设计：赵爽

主要材料：不锈钢管、玻璃、毛皮草、*LED*光源、爵士白大理石、黑白根大理石

摄影：孙东宝、利旭恒

麻辣诱惑在上海又刮了一场麻辣风暴，乱箭穿心狂野令人窒息，不锈钢管环绕着每一个圆卡座，让用餐的人们感受身处都市丛林之中，人心的沉淀需要彼此坦诚相待，卸下心防，回归真实的自我，奉献给对方最真实的自我。

The spicy temptation to in Shanghai has scraped a spicy storm, arrows mandrel wild suffocating, stainless steel tube around the deck every round, so that people feel living in dining among the urban jungle, the people of precipitation need to remove the psychological barriers honest with each other and return to the true self, devotion to each other the most authentic self.

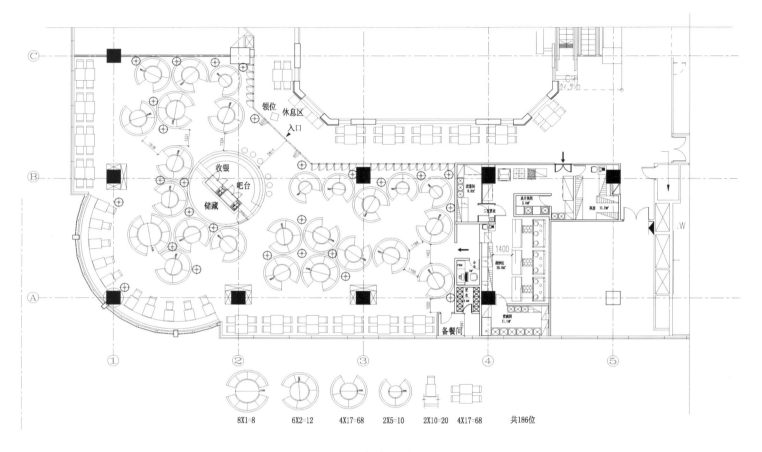

平面图/Plan

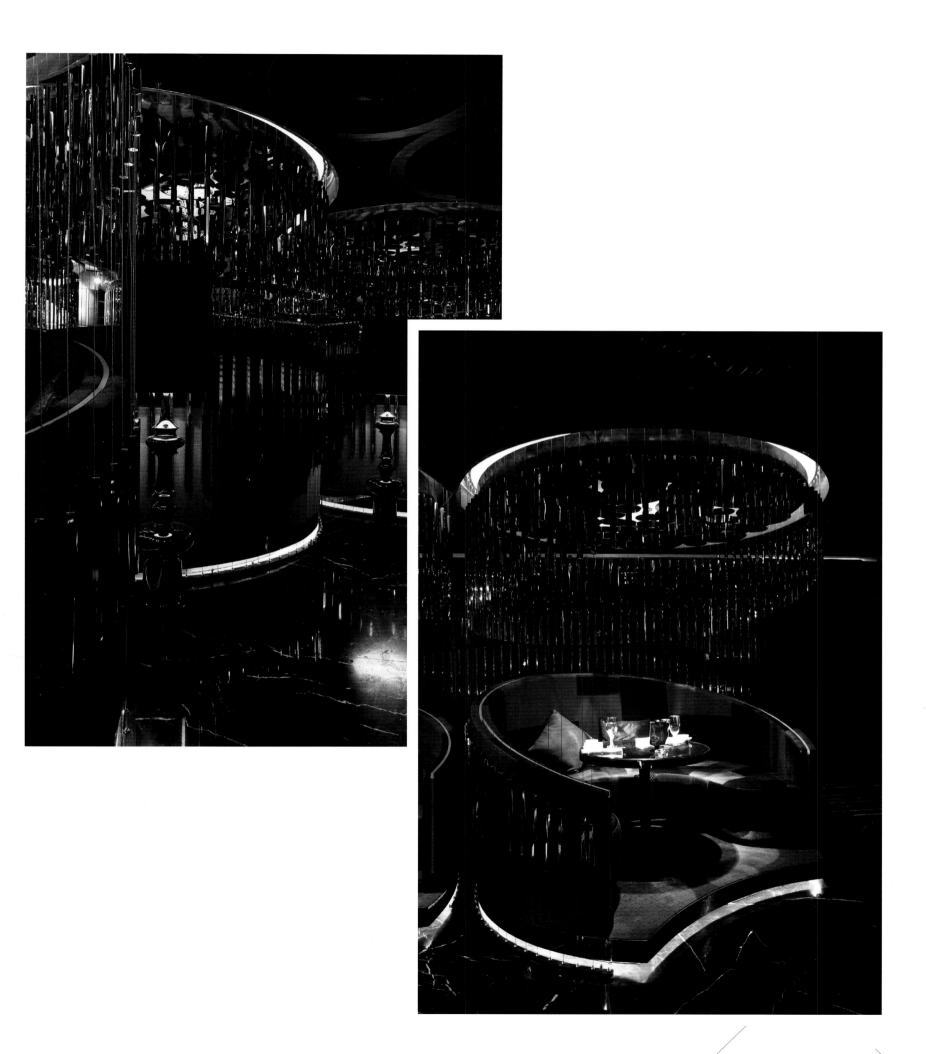

## 利旭恒:

出生于中国台湾，英国伦敦艺术大学 BA (Hons) 荣誉学士。

GOLUCCI DESIGN LIMUTED古鲁奇公司设计总监；长年致力于酒店餐饮空间、专卖店与地产商的设计工作，10多年的酒店餐饮零售空间设计，累积了丰富的项目经验，在多个设计风格与商业形态上都卓有成就。

## 近期主要荣膺:

2009年度金外滩最佳餐厨空间优秀设计奖（美国室内Interior Design）；

2008年度中国十大样板间设计师（中国建筑装饰协会）；十佳配饰设计师（中国第二届）；

2008年，作品鼎鼎香餐厅获北京最佳火锅餐厅称号（City Weekend）；

2007年，作品鼎鼎香餐厅获北京最佳餐厅称号（That's Beijing）；

2006年度中国十佳设计师（中国建筑装饰协会）；

2004年度中国十佳设计师（中国住交会）。

# Spicy Temptation

# 麻辣诱惑

## 淮海中路店

项目面积：*3000m²*

项目地址：上海淮海中路4,5,6层

主创设计：利旭恒

参与设计：赵爽

主要材料：不锈钢管、玻璃、毛皮草、*LED*光源、爵士白大理石、黑白根大理石

摄影：孙翔宇

麻辣诱惑旗舰店坐落于上海著名的淮海中路茂明南路交口。 这里有老上海别致的环境特色及时尚艺术气息，麻辣诱惑旗舰店周遭充斥着各大国际品牌ZARA , H&M, C&A, 吸引了许多潮男潮女及时尚活动在此举办，俨然为上海历久弥新的时尚地标。麻辣诱惑在京、沪两地都是极受欢迎的餐饮连锁品牌，设计师利旭恒由麻辣诱惑品牌LOGO代表女性胴体的诱人曲线中发想，透过诠释时尚女性自信自主的理念，在空间的规划中延续品牌精神并藉此表达现代都会女子理性与感性兼容的特质，及享受社交的生活态度。

Spicy temptation flagship store located in the famous Huaihai Zhong Road, and Maoming South Road intersection, shanghai. There are old Shanghai and unique environmental features and fashion art, spicy temptation of flagship stores around filled with major international brand ZARA, H & M, C & A, has attracted a lot of influx of men and women and fashion events held here, as if to Shanghai remained intact New fashion landmark. Spicy temptations in Beijing and Shanghai are very popular restaurant chain brands, designers, Xu Hang Lee tempted by the hot brand on behalf of women Tong LOGO attractive body hair to the curve, self-confidence of women through the interpretation of the concept of fashion, continuity planning in space and to express the modern spirit of the brand will be compatible with rational and emotional woman character, and enjoy the social attitude towards life.

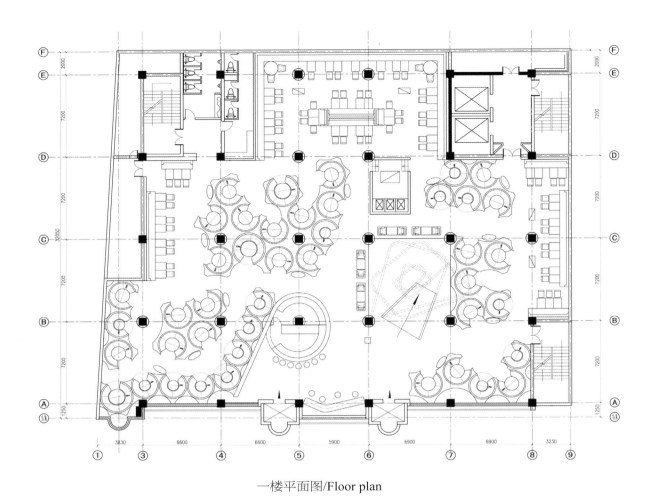

一楼平面图/Floor plan

设计师利旭恒强调，我们刻意传达"空间表情"的概念，这样一个可以让我们尽情放松心情，同享美酒、佳肴、音乐的场所，不论是年轻情侣相约制造浪漫，美食主义者挑剔用餐氛围或是工作同伴犒赏自己…在相同的空间里，藉由不同的人物、背景，交织述说着都会传奇。

Designers communicate "face space" concept, which allows us to enjoy a relaxing mood, share wine, food, music, places, whether it is similar to create a romantic young lovers, food advocates critical of the work of peers or reward dining atmosphere themselves ... in the same space, by different characters and backgrounds, describing the intertwined legend.

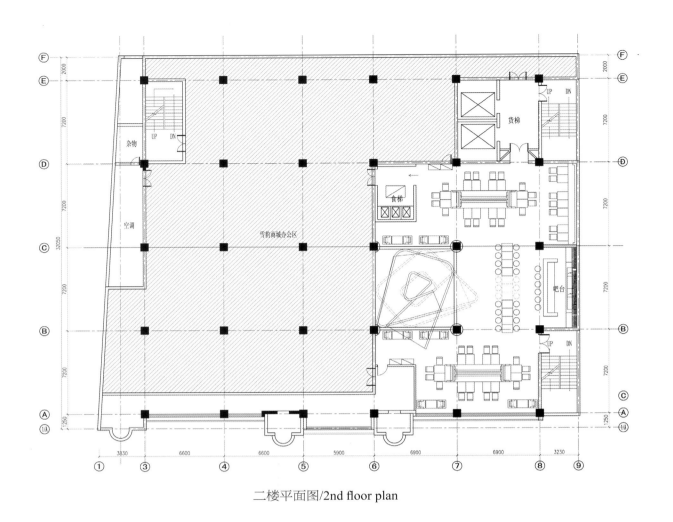

二楼平面图/2nd floor plan

本案在空间操作手法上着重于不等比重的空间分割概念，一座看似复杂且层层交错的空间主体结构，透过三层挑高的空间玻璃墙面与巨大的曲线造型楼梯导入设计概念主题，清晰的引导出这座城堡的空间动线，设计师希望藉此引寓现代都会女子理性与感性兼容的特质，及享受社交的生活态度。

Approach on the case in space operations focused on the concept of varying the proportion of the space division, a seemingly complex and layers of the main structure of the space staggered through the three high-ceilinged space with a huge glass wall staircase curve modeling design concepts into theme , clearly leads to the castle in space moving line, the designer hopes that lead women blending of modern rational and emotional characteristics are compatible, and enjoy the social attitude towards life.

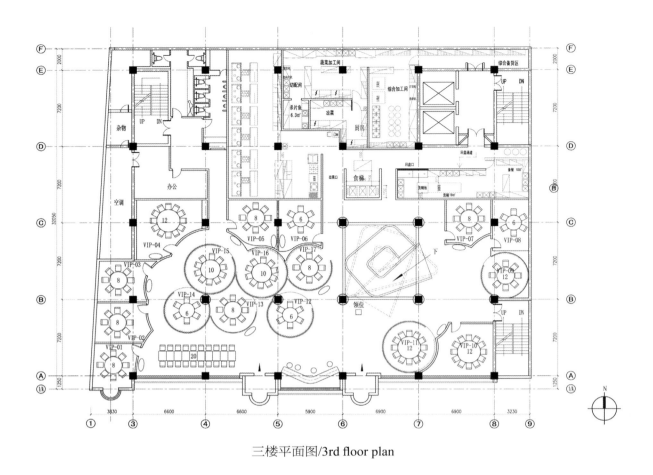

三楼平面图/3rd floor plan

接待区的首层空间由一个个大小不一、高低不同，但是性质统一的圆形构成，圆形圈椅与圆形餐桌与上方的大红圆灯交相呼应，这是设计师特意为了中和强烈的麻辣给人们带来味蕾的快感之时，用视觉的方法来舒缓人们的炙热触感。

The ground floor reception area of space is made of many circular that large small, high or low, but nature of a unified composition, chair and round table round the top of the red circle with cross-lights echo, which is specifically for the design and strong spicy taste buds in bringing pleasure to people when the visual way to ease people's hot touch.

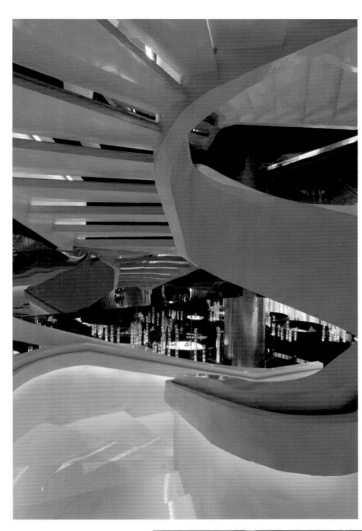

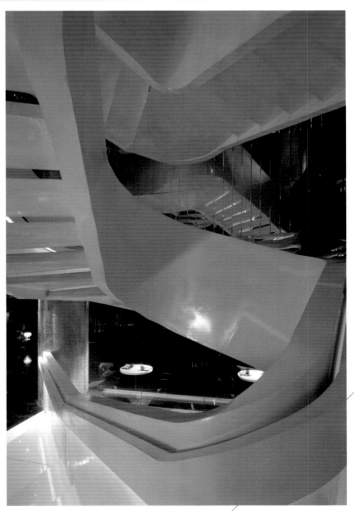

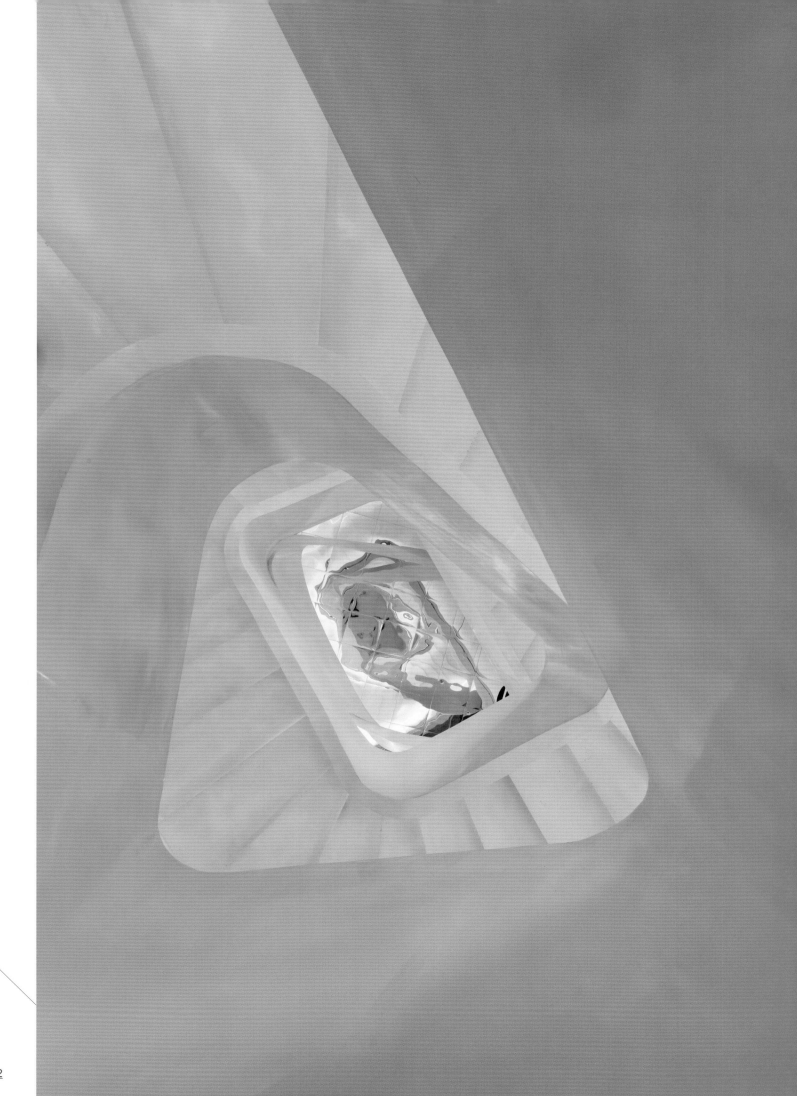

## 刘世尧：

高级室内建筑师；

中国建筑学会室内设计分会第十五专业委员会副主任

美国IFDA国际室内装饰设计协会河南办事处主任；

河南省建筑装饰协会家装委员会副秘书长；

河南鼎合建筑装饰设计工程有限公司执行董事。

## 近期主要荣膺：

2009年，获CIID中国室内设计二十年"优秀设计师"称号；获第十七届亚太区室内设计大奖赛荣誉奖；获中国室内设计大奖赛商业工程类二等奖；获中国风-IAI 2009亚太室内设计邀请赛铜奖。

2008年，获亚太室内设计方案类金奖；获第四届海峡两岸四地室内设计大赛铜奖；获中国室内设计大奖赛酒店、宾馆工程类三等奖；获上海"金外滩"佳作奖。

2007年，获IFI第三届国际室内设计大奖赛优秀奖及佳作奖；获"全国百名优秀室内建筑师"荣誉称号。

2005年，获"全国住宅装饰装修优秀企业家"荣誉称号。

2004年，工程获全国住宅装饰装修示范工程奖。

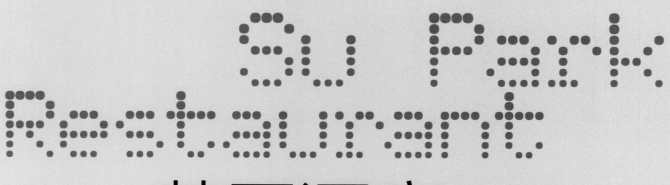

# 苏园酒店

项目面积：*5400m²*

项目地址：郑州市陇海路与交通路交叉口

主案设计师：刘世尧、孙华锋

设计公司：河南大木鼎合建筑装饰设计工程有限公司

竣工时间：*2010年1月*

　"风雅吴地、水墨江南"，留下无数文人墨客的行迹，也留下人们印象中一幅浓淡相宜的水墨长卷，这就是我们在苏园酒店设计中所要传达的思想和意境。

　我们在原有建筑物的加建及改建上，植入了中国传统的"院落"形式作为设计的理念。首先在平面规划上，把苏州园林的"园"、"巷"的意境体现在酒店的中庭庭院和大堂的入口处，既营造了灵透的入口中庭的空间，又很自然的把江南建筑的元素和特质体现出来。

"Wu civilization, southern culture," leaving many scholars trace the line, also left the impression that people in an ink scroll, this is what we in the hotel design to convey the thoughts and mood.
We are in the original building additions and alterations, the implantation of a traditional Chinese "courtyard" in the form as a design concept. First, planning in the plane, the Suzhou gardens of the "Garden", "alley" in the mood reflected in the hotel's courtyard and the entrance lobby, both through the entrance to create a spiritual space, and very natural elements and the South China Architecture characteristics reflected.

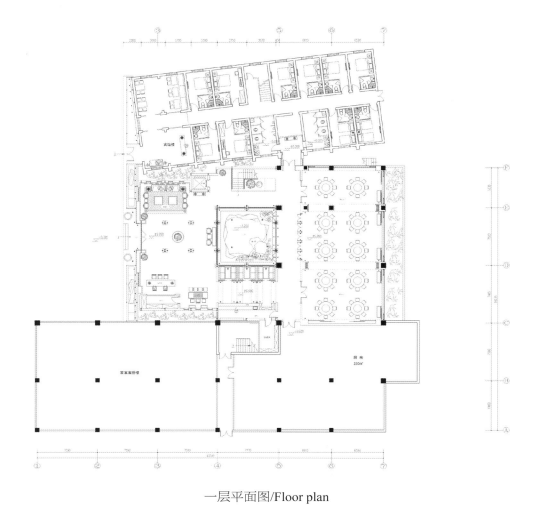

一层平面图/Floor plan

外立面向内单坡的屋顶，既体现江南建筑的灵秀特征，又体现中国民居中常用的"四水归堂"的手法，入口处大片对称的实墙和中庭对外大片洁净的玻璃形成强烈的虚实对比，变异简洁的中式窗格在外立面中细致动人的分格、对称的入口格局使得整个建筑既大气又精致。

Facade of the roof, not only embodies the characteristics of the building south, but also reflects the approach commonly used in Chinese houses, at the entrance to a large real symmetry of external walls and the large clean glass court in sharp contrast, simple Chinese-style facades in the pane in detailed sub-moving grid, symmetrical pattern of the entrance and refined atmosphere make the whole building.

围坐在中庭、走廊的客人，透过窗格、竹帘、纱幔，似触非触，这种环境下的闲情雅致，让身在其中的每一位客人不禁赏心悦目、心旷神怡，顾客络绎不绝。

Sitting in the atrium, corridors guests, through the pane, bamboo curtain, feeling elegant leisure in this environment, so that the body in which each guest can not help but pleasant, relaxed and happy, of patrons.

环绕入口中庭及景观庭院的内侧，我们大面积的采用深色变异简化了的实木窗格，简朴而别具一格的造型，使得内外空间互相渗透。采用黑灰色的柱子、白色的墙体，及简化抽象的中式门窗，既简洁又明快的体现水墨江南文化的精髓。中庭的上空悬挂着高低错落的渔网及鸟笼，不但丰富了空间，而且也充分体现了江南的鱼米之乡和悠闲的市井文化。

And landscape around the entrance atrium inside the courtyard, we used large dark wood pane, create a simple form, making the space inside and outside the mutual penetration. We use black and gray columns, white walls, and simplified Chinese abstract windows and doors, the embodiment of both concise and lively essence of southern culture. In the fishing nets hanging over the court and the cage, not only enriched the space, but also fully embodies the rich South and relaxed public culture.

在包房设计中，我们将江南富有、水乡诗一般的画面，通过不同的材质、手法去展现。从家具的设计到室内的陈设，都力求简约明快又不失大气，呈现出温馨、典雅、舒适、厚重的空间效果。既符合中国人所崇尚的人文环境，又通过对中国传统的理解，把江南文化的理念彻底体现。

In the private room design, the south of rich, poetic images of rivers and lakes, through different materials, techniques to show. From furniture design and interior furnishings, have sought to simple yet lively atmosphere, showing a warm, elegant, comfortable, thick space effect. Consistent with respect for the Chinese people's cultural environment, but also through the traditional understanding of the concept of the southern culture completely reflected.

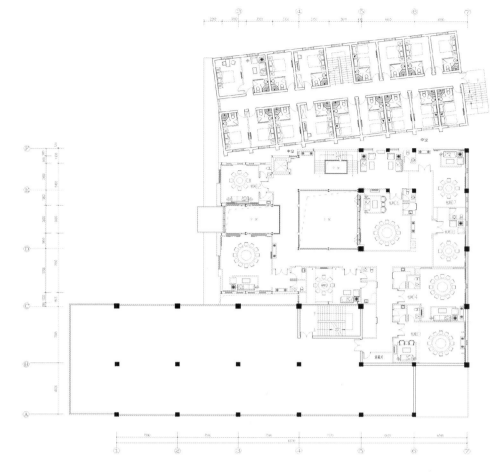

二层平面图/2nd floor plan

## 利旭恒：

出生于中国台湾，英国伦敦艺术大学 BA (Hons) 荣誉学士。

GOLUCCI DESIGN LIMUTED古鲁奇公司设计总监；长年致力于酒店餐饮空间、专卖店与地产商的设计工作，10多年的酒店餐饮零售空间设计，累积了丰富的项目经验，在多个设计风格与商业形态上都卓有成就。

## 近期主要荣膺：

2009年度金外滩最佳餐厨空间优秀设计奖（美国室内Interior Design）；

2008年度中国十大样板间设计师（中国建筑装饰协会）；十佳配饰设计师（中国第二届）；

2008年，作品鼎鼎香餐厅获北京最佳火锅餐厅称号（City Weekend）；

2007年，作品鼎鼎香餐厅获北京最佳餐厅称号（That's Beijing）；

2006年度中国十佳设计师（中国建筑装饰协会）；

2004年度中国十佳设计师（中国住交会）。

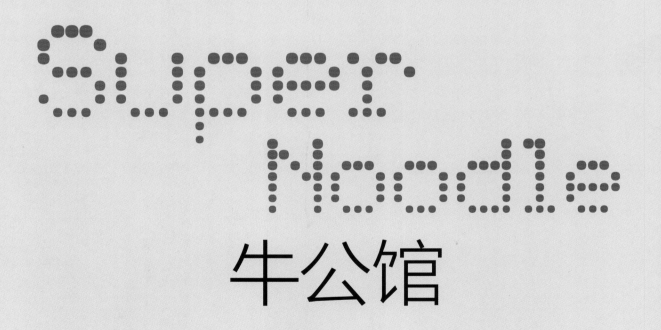

# Super Noodle
# 牛公馆

项目面积：*250m²*

项目地址：北京朝阳区东三环佳程广场*7*层

主创设计：利旭恒

参与设计：赵爽

主要材料：青花瓷大碗、筷子、茶色镜、橡木、*LED*光源、中国黑大理石

摄影：孙翔宇

牛公馆是一家传统的"台湾川味牛肉面馆"，在台湾成长的孩子都有共同的记忆……，昏黄的灯光，满屋花椒大料与中药材飘香，大大的青花瓷面碗，一桶桶的筷子满桌，滚烫的热汤浮满了黄黄的牛油，能够塞满整嘴的大块牛肉，大口咬下白粗弹牙的面条……。20多年前的回忆了，那年老张已经70岁，四川绵阳人，1949年跟随国民党部队到台湾，退伍后离家40年的老张用着对故乡残存的记忆拼凑出属于他家乡的味道，这也就是今天的台湾川味牛肉面。

Children growing up in Taiwan have a common memory of ... ..., dim lighting, room full of aniseed and pepper Chinese herbal fragrance, big blue and white porcelain bowl, chopsticks barrels full table, hot soup full of yellow butter floating, to the whole mouth full of chunks of beef, big mouth and thick shells of white teeth bite of noodles ... ..... Zhang used the remnants of the home belonging to his memory to piece together the taste of home, which is today's Sichuan beef noodles in Taiwan.

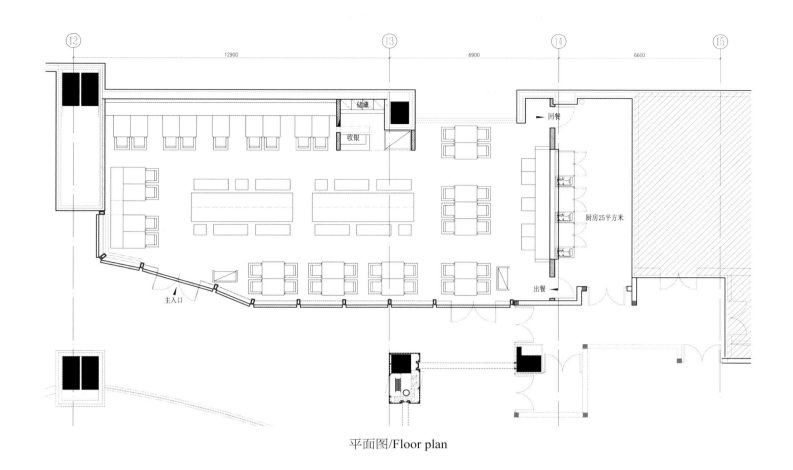

平面图/Floor plan

当时年轻的我怎么会了解一个离家多年，外省四川老兵的心理，好吃是每回那一大碗一大碗的面带给我最满足的回忆。当2010年在北京时，一个台湾客户委托我设计一个牛肉面馆，霎那之间那年吃面的回忆有如顷巢而出，同时也开始了我的筑梦工程… 一个充满吃面回忆的面馆。

I do not know at that time a young veteran of the psychological, a delicious bowl of noodles every time it gives me memories. When in Beijing in 2010, a client asked me to design a Taiwanese beef noodle, instant noodles that year between the memories are out of the nest, also began to build my dream project ... a face full of memories of eating noodle.

**姜湘岳：**

高级室内建筑师；

江苏海岳酒店设计顾问有限公司设计总监。

**近期主要荣膺：**

首届中国最具商业价值餐饮娱乐类设计 全国50强；

中国风-IAI2009亚太室内设计精英邀请赛 三等奖；

**2009 INTERIOR DESIGN CHINA**酒店设计奖 优秀奖；

2009年度中国饭店业设计装饰大赛-金堂奖 酒店餐厅类银奖；

2009年度中国饭店业设计装饰大赛-金堂奖 中餐厅类铜奖；

2009年度中国饭店业设计装饰大赛-金堂奖 中国十大餐厅空间设计师；

1989-2009中国室内设计二十年 优秀设计师；

2010年中国（上海）国际建筑及室内设计节 "金外滩奖" 最佳酒店设计奖（优秀）；

2009-2010年度室内设计百强人物；

第4批全国有成就的资深室内建筑师称号。

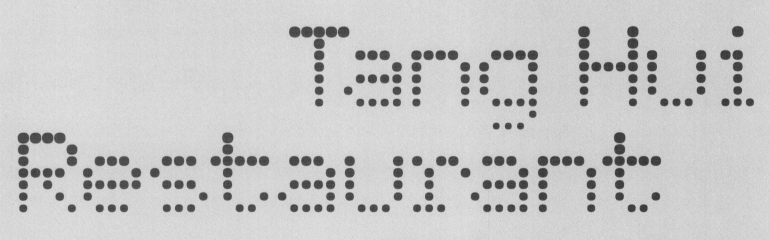

# Tang Hui Restaurant

# 唐会

项目地址：南京西家大塘46号

设计师：姜湘岳

设计公司：江苏省海岳酒店设计顾问有限公司

这是一个环绕在优美风景周围的餐饮建筑——唐会。从名字听起来就感觉带着中式古典宫廷的元素，我们的设计也就从此展开。

我们在设计过程中，首先对这个建筑进行了整形，使它最终能够内外合璧。设计中，我们将中国古典主义建筑和现代主义相结合、虚和实相结合，专门为这个餐饮设计的标识，在白色的墙壁上显得特别醒目。通透的室内将建筑整体的表现出来，并把室内温暖的感觉传递到室外来。夜晚迷人的背景状态，有如一场戏剧拉开了序幕。

It is a catering building surrounded by the beautiful sceneries – Tang Hui. It sounds like an element with Chinese style classic palace. Our design is just developed from this point.
During the design, we have re-shaped this building and made it achieve its final internal and external integration. In this design, we have combined the Chinese classism building and modernism as well as falsehood and reality together to specially design a symbol for this restaurant, which are especially striking on the white wall. The transparent interior has exhibited the entire building and delivered the interior warm feelings to the exterior. The charming background at night is just like a prelude to a drama.

接待大堂迎宾处，我们特意精心挑选了迎宾花与背后的背景形成非常好的对比状态。入口的精致表现，顶和地之间的大体块的结构感，接待台背后存放的古书对空间进行了柔化。走道采用古典主义矮墙窗棂的方式，用镜面加木格的做法，使客人能够在木格的序列中看到周围的影像，仿佛走在苏州园林里的感觉。

At the guest reception area of the lobby, we have deliberately and elaborately selected the guest greeting flowers and the behind background to form a quite good contrast. The delicate demonstration at the entrance, the structural sense of the large blocks between the ceiling and ground as well as the ancient books stored behind the reception desk has softened the space. The corridor has applied the mirror together with wood grids in a way of the classism low-wall muntin so that the guests can see the surrounding images from the serial wooden grids as if they are walking in a garden of Suzhou.

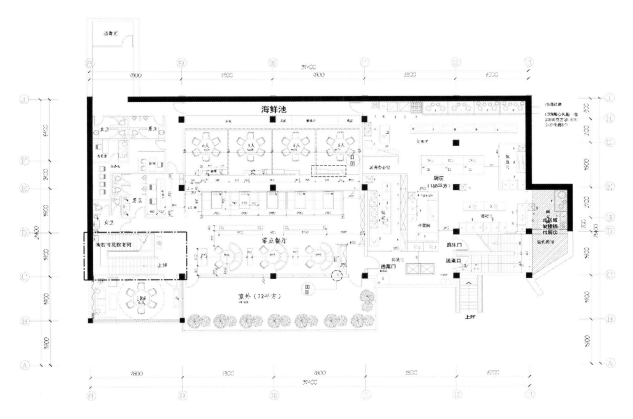

一楼平面图/Floor plan

豪华包间中，我们把书香门第的气氛引进来。在餐厅里面也放置了大量的古典主义书籍，使得空间充满了浓浓的书香气味。我们从坡屋顶得到的灵感而设计的吊灯和深色的木饰面以及墙面上的仕女图形成一种古典主义高贵的感觉。

In the luxurious VIP rooms, we introduced the literary atmosphere. The large quantities of classism books are also placed in the restaurant so that the space is filled with the strong book fragrance. The pedant lamps designed through the inspiration we obtain from the sloping roof, the dark wood veneers as well as the beauties' pictures on the wall have displayed a classism and noble feeling.

**陈彬：**

武汉理工大学艺术设计学院副教授

中国建筑学会室内设计分会（CIID）会员

中国建筑装饰协会会员及设计委员会委员

国际室内建筑师/设计师联盟（IFI）专业会员

亚太建筑师与室内设计师联盟（IAI）理事

大木（湖北）后象设计顾问机构 设计主持

## 近期主要荣膺：

作品入选ANDREW MARTIN 2010年度国际室内设计大奖；

2010年，亚太室内设计双年大奖赛-评审团特别大奖；

2010年度，广州国际设计周"金堂奖" 年度十佳餐饮空间设计；

2009年度，APIDA第十七届亚太区室内设计大奖银奖；

2009年度，作品入选德国iF2009中国设计大奖；

2009年度，广州国际设计周"金羊奖"中国十大室内设计师；

2009年度，广州国际设计周"金堂奖"餐饮酒吧类中国十大设计师；

2009年度，Interior Design China "酒店餐厅类最佳设计" 奖；

2009年度，金指环全球室内设计大赛优秀奖。

# The Crimson Restaurant

# 红馆餐厅

项目面积：692m²

项目地址：武汉花园道

完成设计时间：2009年11月

设计公司：大木设计中国(湖北)后象设计顾问机构

主持设计师：陈彬

参与设计师：李健、傅晟、严小兵

摄影：吴辉

感知熟悉的文化表象：简约的现代家具、优雅的古典穹顶、神秘的波斯花纹、前卫的数码图象的奇特组合。

东方、西方、古典、现代、传承、创新，在这个空间中都得到全新的诠释。马赛克呈现传统的地毯、玻璃丝印数码图案……设计师赋予了材料新的使命，也在创造中演绎了全新的文化内涵。

Red House brings you familiar cultural imagery: A unique combination of sketchy modern furniture, elegant and classical castle dome, mysterious Persian pattern, and avant-garde digital pattern.

Red House offers you a brand new interpretation of the east, the west, the classical, the modern, as well as the inheritance and innovation, such as traditional carpets with Mosaic display, glass printed digital images···, to which the designer has endowed material with new mission,with the creation of new cultural connotations.

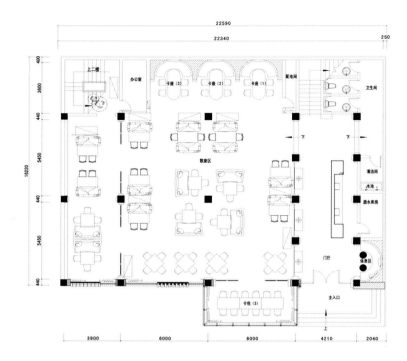

一层平面图/Floor Plan

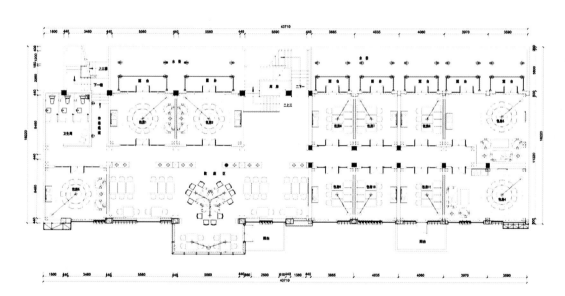

二层平面图/2nd Floor Plan

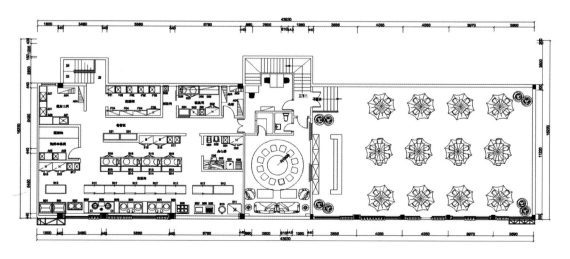

三层平面图/3rd Floor Plan

李智：

　　毕业于广州美术学院设计系环境艺术设计专业；法国国立艺术学院设计管理硕士。

　　现任：广东省装饰总公司设计总监。

　　全国有成就的资深室内建筑师；

　　全国有成就的高级环境艺术设计师；

　　法国设计师《FNSAI》协会会员；

　　中国建筑装饰协会设计委员会（企业部）副主任委员；

　　中国建筑协会室内设计分会广州专业委员会委员；

　　全国建筑与室内设计师俱乐部领导小组广东联络处执行委员；

　　IFDA国际室内装饰设计协会会员；

　　广东省工商联设计师工会理事；

　　广东省装饰行业协会设计师专业委员会专家；

　　广州市建筑装饰协会设计委员会副主任；

　　广州日报设计师顾问团成员；

　　《IN-D》杂志编委；

　　广东省艺术设计高级专业资格评审委员会委员；

　　广东省艺术设计中级专业资格评审委员会委员。

# Tomidagiku Restaurant

# 富田菊日本皇尚料理

项目面积：*3000m²*
项目地址：广州
主案设计：李智

富田菊是以铁板烧及和食为主的日式餐厅，面积约3000平方米。东方的文化神韵在本源上有着许多的共性，相比之下日式餐厅更注重环境气氛上的传达，因此在风格的定位上采用了现代和式的装饰手法，运用当代的装饰材料与天然建材再现传统造型的文化图藤，通过环境灯光的烘托，营造出一个富有鲜明现代和式风格的就餐气氛，整体线条一气呵成，细微处适宜的曲折变化，在虚实与有无之间给人留下无限的遐想空间。

Case is the Japanese restaurant, an area of about 3,000 square meters. Eastern culture has a lot of charm on the common origin, compared to Japanese restaurants pay more attention to communication on the atmosphere, so in the style of positioning and style with modern decorative techniques, the use of contemporary decorative materials and natural building materials reproduction of traditional cultural map style rattan, by contrast lighting environments, creating a rich distinctive style dining environment. The whole line is smooth, subtle twists and turns the appropriate changes in the actual situation and the availability of people between the infinite imagination of space left.

沿着富有趣味性的小桥，漫步而
至，营业大厅豁然开朗，透过墙
体的装饰和气氛的营造，点题性
的"菊"在这里充分得到了体验。
包房、卡座设于走廊的两侧，功能
与装饰虽各有不同但风格统一，
在设计的实施中通过几大材质与
色彩坚定把握，以及各设计元素
的分量、比例表现位置与时机的
控制，达到了简练而不失高雅，形
象价值得到了提高。

Along the bridge into the
business hall, wall decor
and atmosphere through the
building, chose this topic of
"Chrysanthemum" was fully
experience here. Private dining
room, deck located on both
sides of the corridor, functional
and decorative styles are quite
different, but uniform in the
design of the implementation
of several materials and colors
through a firm grasp, and the
weight of the design elements,
location and timing of the
performance ratio control, to
the concise and yet elegant,
the image of value has been
enhanced.

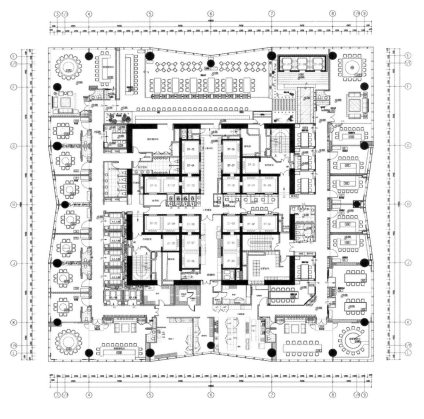

平面图/Plan

在平面的功能设计上，餐厅的营业区主要分布于建筑的三个朝向，无论是大厅还是房间的食客在品尝美食之余都能透过巨型的外墙玻璃把世界级建筑大师的作品尽收眼前，味觉与视线得到充分的享受，身处其中，无形的价值瞬间再一次获得提升。Function in the plane design, the restaurant is open mainly distributed towards the construction of three, either the hall or room, diners, I can taste the food of the giant wall of glass through the world-class architectural works best closed eyes, taste and sight to the full enjoyment.

## 朱明：

职业空间艺术设计师

朱成私立石刻博物馆 副馆长

新西兰巴码设计试验工作室 设计总监

新西兰AUT 艺术学院 空间设计系（Spatial Design）硕士

新西兰设计师协会 会员

成都石筑营造环境艺术设计有限公司 董事长

## 近期主要荣膺：

2005年，设计作品"Datascape"被AUT空间设计专业和奥克兰大学建筑系采用为教学范例；

2006年，设计作品"I-Bar" 参加奥克兰第一届建筑艺术节，并录入Barcode丛书；

2006年，与Yosop Ryoo（韩）在新西兰注册成立"巴码空间设计实验室"（Barcode Design Laboratory LTD., New Zealand）并担任其设计总监；

2007年，设计作品"Wavegarden"参加奥克兰Viaduct港口艺术节；

2007年，设计作品"I-Bar"获新西兰年度最佳设计奖（Best Awards），获奖作品在全国进行巡展（Best Awards是新西兰每年举行的集平面设计，工业设计，空间设计和建筑设计的综合类重大设计比赛）；

2008年，参与"成都宽窄巷子片区改造"并独立负责其部分空间设计项目；

2008年，独立负责并完成艺术家朱成、王亥、谭云于清河艺术家村的工作室建筑设计。

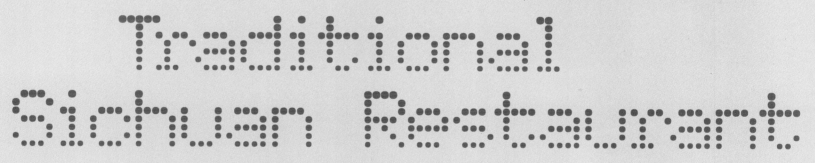

# Traditional Siohuan Restaurant
# 南城记川菜馆

项目面积：*700m²*

项目地址：四川省成都市上池正街28号

设计公司：成都石筑营造环境艺术设计有限公司

项目主持：王亥

设计总监：朱明

摄影：贾方

位于城南，因南得名，诗化，谓之南城记。"记"之含义：一为商号；二为记忆，旨在以十个故事告慰那些梦中残存的记忆，也可称之为我们那个成长年代的记忆。我们试着以自己的方式诠释餐饮，也试着以一种别样的设计理念来解释饮食与空间的关联。更想以一种方式来经营意大利设计产品，其吃、喝、坐、观、用，可合而为一，转化为一综合空间——Food & Home（食与家）。

Located south, named for the South, so named for the South City in mind. "mind " of meaning: one for the store; the other for the memory to recall the story, we can call it that grew up in memory. We try to explain the restaurant in its own way, also tried a different kind of design concept to explain the food and space. Among them, more like in a way to run the Italian design products, eating, drinking, sitting, view, use can be combined into one, into an integrated space - Food & Home.

约半年前我和南城记投资人有个约定，做一间不一样的餐馆。其宗旨，老老实实做人，认认真真做菜。或者说，认认真真做人，老老实实做菜。其人其菜，同出一道。正如我（王亥）十二年前香港做的那间餐馆一样，"文如其人"，餐馆如其人。我把这间餐馆称作"作者餐馆"。其后，被人俗不可耐的改来改去，俗称"私房菜"。

Restaurant investors want to design a different restaurant. Mission: honest man, seriously cook. His life, his food, with a one. As I did 12 years ago in Hong Kong, like restaurants, "Style is the Man", the restaurant as his face. I called this restaurant "of the restaurant. " Subsequently, was changed to "private the food. "

我问投资人这间餐馆做大了能不能做到这些？或者自问，换个地方，换个"博物馆"，换个"画廊"，换个"意大利设计空间"吃一顿饭将会怎样？这是一个素色，可自由处置的建筑空间，尽用一个在餐馆空间中可能发生的意大利设计，将会怎样？我想，一定是个有趣的经验。

I asked the investors can design this restaurant as a "museum"or"gallery"or "Italian design space? " If this location, the restaurant's design can be freely disposed of, the design of the restaurant must be a fun experience.

地方是换了，服务要如一，出品怎么办？在如此空间内吃出一个"苍蝇馆"的味道（生就喜好"苍蝇馆"的味道），将该怎样？这是一个突发的想象。这个约定就是"南城记"。回到前说，老老实实做人认认真真做菜，这是一个问题。或者，古人所曰："形而上，形而下"，其实是个"道""器"之谓，"饮食男女"寻常事，其实也是个做人的道理，做餐馆的道理。其实这是个艺术家的想法。谁叫你是艺术家出生？跑十二年堂，且做设计，加之策划。这一定是个"另类"。这就是等待你来参与的"南城记"。

Place is changed, the service should be as one, producing or how to do? This convention is the "South City in mind. " Back before that, an honest man, seriously cook. In fact, the important thing is "Tao" and "device","diet " is a very unusual thing, in fact, also a man of reason, reason to do restaurants. In fact, this is the artist's idea. Who told you artists?This is waiting for you to participate in the "South City in mind. " The design of the restaurant is the artist's "alternative" work.

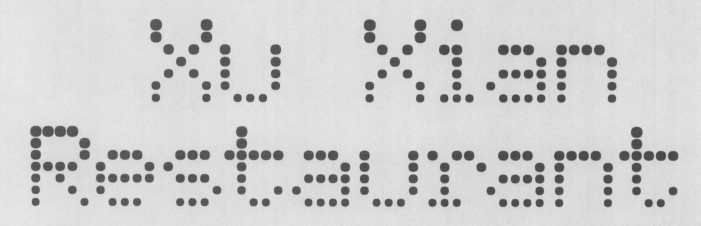

# Xu Xian Restaurant

# 许仙楼

项目面积：*3000m²*
项目地址：北京朝阳区工人体育场西门内
概念设计：余青山
主持设计：徐宗
执行设计：韩宇、张晓燕
主要材料：马来漆、石材、木地板

似乎是一夜之间，工体西门冒出来一个"许仙楼"，来势汹汹势不可挡——开业当天人都快把门给踏破了，且来捧场的多是本城文艺界大腕。许仙楼名字够飘逸，更惹人联想起许仙的媳妇儿白娘子。可这儿没有白娘子，也不卖药救人，单单出品精致杭帮大菜，以粤菜为辅。

Seems to be overnight, Simon came out of a public body, "Xu Xian Restaurant", the raging unstoppable - the opening day of people to wear out the door faster, and to join the biggest names in many literary and art circles of this city. The name of enough to elegant floor, but also catches the young married woman recalls White Snake. White Snake can not here, not sell drugs to save people, produced just to help fine-Hangzhou dishes to cantonese supplemented.

四层独栋的白色别墅式餐厅堪称超大手笔了。餐厅门前豁然开朗两片波光粼粼的水域，很有现代范儿的海市蜃楼感，落地大窗明亮透彻，每层都有露台伸出楼外，气质清雅脱俗。

Four single-family style restaurant called the white house a super generous. Restaurants suddenly see the light in front of the sparkling waters of two pieces, very modern sense of Fan mirage of children, bright and thoroughly French window, each floor has a terrace out outside the building, temperament elegant and refined.

一层散台，正中心的水道设计迎合室外水域；二层十间包房，尤其推荐南侧阳光玻璃大包，在西式范儿的空间里享受中式江南细节；三层四间贵宾包房，商务宴请十分有面儿。

Layer of scattered units in the center of the channel design to meet outdoor water; the second floor rooms, especially the south side of the sun glass large package recommended to enjoy the Chinese the details; three-four VIP rooms, ideal for business dinners.

主厨是特地从杭州某传统名店请来的，一点一点地挪移来杭帮美味，整体的口感也都保持传统的淡雅路线。

Chef is a specially chef invited from Hangzhou, a famous restaurant of traditional, authentic Hangzhou cuisine, also to maintain the overall taste of the elegant traditional route.

## 利旭恒:

出生于中国台湾，英国伦敦艺术大学 BA (Hons) 荣誉学士。

GOLUCCI DESIGN LIMUTED古鲁奇公司设计总监；长年致力于酒店餐饮空间、专卖店与地产商的设计工作，10多年的酒店餐饮零售空间设计，累积了丰富的项目经验，在多个设计风格与商业形态上都卓有成就。

## 近期主要荣膺:

2009年度金外滩最佳餐厨空间优秀设计奖（美国室内Interior Design）；

2008年度中国十大样板间设计师（中国建筑装饰协会）；十佳配饰设计师（中国第二届）；

2008年，作品鼎鼎香餐厅获北京最佳火锅餐厅称号（City Weekend）；

2007年，作品鼎鼎香餐厅获北京最佳餐厅称号（That's Beijing）；

2006年度中国十佳设计师（中国建筑装饰协会）；

2004年度中国十佳设计师（中国住交会）。

# 烹大师烧肉达人

项目面积：**300m²**

项目地址：上海四川北路嘉杰广场**4**层

主创设计：利旭恒

参与设计：赵爽

主要材料：木炭、钢管、玻璃、*LED*光源、花岗毛石、灰黑色地砖

摄影：孙翔宇

"木炭"净化空气同时净化人心。

设计师利旭恒以烧肉的主要燃料"木炭"为设计主题的烧肉店，享尽奢华过后的人们需要回归到原点让心灵自我沉淀，用一个自然根本态度享用美食。餐厅环境中再也没有任何多余的装饰，取而代之的是料理烧肉的燃料，经过设计师分割的木炭块状墙体，利用人们丢弃的碎杂木组合而成了餐厅大门，天花板则是延续了木炭块状墙体的分割，设计师希望以自然环保的绿色设计，带给人们对环境的重视。

"Charcoal" clean air while purifying the people.

The main fuel for the designer to roasted pork, "charcoal" as the design theme. People need to enjoy the luxury after return to the origin, with the natural attitude meal. Restaurant environment, there is no unnecessary decoration, after the designer of the charcoal block partition walls, hardwood pieces discarded by the people into the restaurant door, and so the combination, designers hope to the natural environment of green design, to bring people the environment seriously.

餐厅是以经营肥牛为主的火锅料理。空间分三层，整体运用不多的材料种类以追求用餐环境应有的舒适度，并结合不同的空间处理，保证三层分别以各自的感受展示给客人，配合合理的灯光及艺术和自然的大体积装饰品，以使人产生冥想和回忆。

Restaurant is a pot with main dishes beef. Space divided into three layers, the overall use of the few types of materials due to the pursuit of comfort dining environment, combined with different spatial processing, to ensure the feeling of three-tier display of each of their guests, with reasonable lighting and art and nature. The large volume of decorations, to engender meditation and memory.

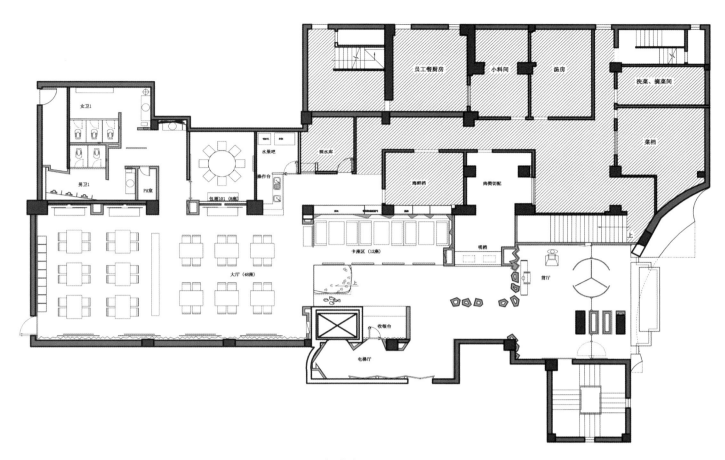

平面图/Floor plan

## 林文格：

L&A文格空间设计顾问公司创办人，创意总监；

高级室内建筑师；

全国百名优秀室内建筑师；

IFDA国际室内装饰设计协会理事；

ICAD国际A级职业景观设计师；

中国建筑学会室内设计分会第三专业委员会委员；

香港室内设计协会中国深圳代表处委员；

意大利米兰理工设计学院室内设计硕士。

## 近期主要荣膺：

2010年，英国Andrew Martin 设计大奖；

2009年，美国Hospitality Design Awards 酒店空间设计大赛最高荣誉winner大奖；

2009年，荣获中国建筑学会"中国室内设计二十年二十人"荣誉称号；

2009年，荣获中国建筑学会"中国杰出室内建筑师"荣誉称号；

2009世界酒店"五洲钻石奖"-"最佳设计师"；

APIDA第11届亚太区室内设计大赛餐馆酒吧类别冠军奖；

APIDA第十四届亚太区室内设计大赛酒店类别铜奖。

# 中森名菜
## 海岸城店

项目面积：*16000m²*

项目地址：深圳市南山区海岸城

设计公司：文格空间·设计顾问（深圳）公司

设计师：林文格

主要材料：锈石、毛面大理石、埃比蒙木、墙纸、镜钢、艺术玻璃、澳洲砂岩

本案定位为现代日本风格餐厅。在日本传统文化中，选取樱花、清酒、纸灯笼等最具代表性的直观印象符号，演化为装置艺术品；采用原始古朴材料；借鉴日本传统造景技艺以及对面积利用最大化的考量方式。通过夸张、变形、简化、模拟等多样化的处理手法，将古典日式元素融入时尚商业空间。

Positioning the case for the modern Japanese style restaurant. In Japanese traditional culture, select cherry, sake, paper lanterns and other visual impression of the most representative symbols, evolved into installation art; using original ancient materials; learn from Japanese traditional landscaping techniques and considerations to maximize the use of the way area. Through as a variety of processing techniques, exaggeration, distortion, simplification, simulation, the classical Japanese elements into the fashion business space.

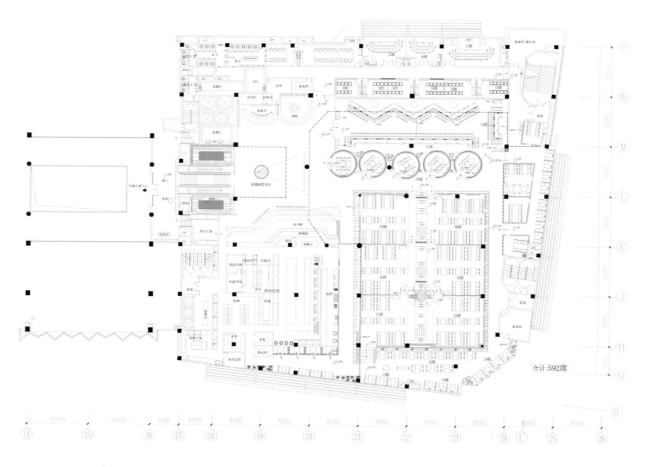

平面图/Floor plan

项目中有一块抬高区域，顶部是巨大的透光中庭。设计师巧妙地将劣势转化为优势，在突起部分营造人工园林，构成中庭景观。另外，设计师充分利用过渡区域，在大厅铁板烧烤台之间加设雅座，由日式纸灯笼演变而来的巨大灯柱从天花垂下。

Projects have a raised area, at the top of the atrium is a great light. Designers skillfully made disadvantages into advantages in the protruding part of creating artificial garden, landscape composition of the Court. In addition, designers take full advantage of the transition area between China and Taiwan in the hall addition of iron grill lounge, by the Japanese paper lanterns evolved from the huge lamp posts, hanging from the ceiling.

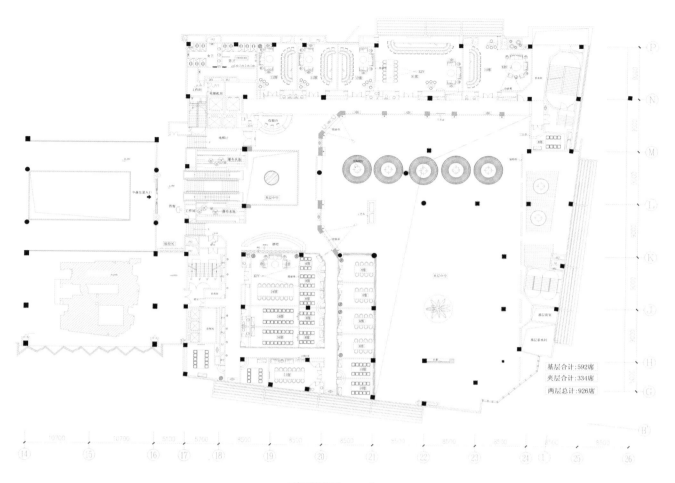

平面图/Floor plan

基层合计:592席
夹层合计:334席
两层总计:926席

设计师通过对整个餐饮空间的解构，形成散座区、雅间区、铁板烧烤区、榻榻米包间、隐私包间等五种就餐模式，提供全开放、半开放、全封闭三种就餐环境，配合相应收费标准，人为创造出极具市场吸引力的超大空间。

Designers through the deconstruction of the entire restaurant space, forming loose seat area, dining rooms areas, iron barbecue area, tatami rooms, private dining rooms in five different modes to provide full open, half open, fully enclosed dining environment, With the corresponding charges, artificially create a highly attractive market, the large space.